Acupuncture and Moxibustion

A Clinical Series

Stroke Rehabilitation

Depression

Headache

Asthma

Herpes Zoster

Knee Osteoarthritis

Lumbago

Dysmenorrhoea

Obesity

Beauty and Skin Care

Insomnia

Perimenopausal Syndrome

Primary Trigeminal Neuralgia

Acupuncture and Moxibustion for

Knee Osteoarthritis

A Clinical Series

Project Editors: **Liu Ying, Shen Cheng-ling, Liu Shui**
Copy Editor: **Zhang Wei**
Book & Cover Designer: **Yin Yan**
Typesetter: **He Mei-ling**

Acupuncture and Moxibustion for **Knee Osteoarthritis**

A Clinical Series

Li Wan-yao

Professor and Doctoral Supervisor

Guangzhou University of Chinese Medicine

Li Wan-shan

Associate Professor

Guangdong Hospital of Traditional Chinese Medicine

Translated by

Hu Yun

Doctoral Degree Candidate of Guangzhou University of Chinese Medicine

Edited by

Nicholas M Dore

Doctor of Oriental Medicine, School of Complementary Medicine

Beijing Herbal Medicine & Acupuncture Inst.

Contributors

Wang Qin-yu & Xing Bing-feng

PMPH 人民卫生出版社

PEOPLE'S MEDICAL PUBLISHING HOUSE

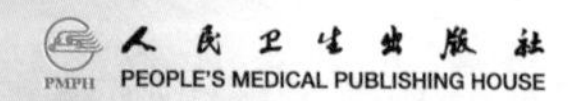

Website: **http://www.pmph.com/en**
Book Title: **Acupuncture and Moxibustion for Knee Osteoarthritis, A Clinical Series**
针灸治疗膝骨关节炎——临床系列丛书

Contact address: No. 19, Pan Jia Yuan Nan Li, Chaoyang District, Beijing 100021, P.R. China, phone/fax: 8610 5978 7399, E-mail: pmph@pmph.com

For text and trade sales, as well as review copy enquiries, please contact PMPH at pmphsales@gmail.com

Disclaimer

This book is for educational and reference purposes only. In view of the possibility of human error or changes in medical science, the author, editor, publisher and any other party involved in the publication of this work do not guarantee that the information contained herein is in any respect accurate or complete. The medicinal therapies and treatment techniques presented in this book are provided for the purpose of reference only. If readers wish to attempt any of the techniques or utilize any of the medicinal therapies contained in this book, the publisher assumes no responsibility for any such actions. It is the responsibility of the readers to understand and adhere to local laws and regulations concerning the practice of these techniques and methods. The authors, editors and publishers disclaim all responsibility for any liability, loss, injury, or damage incurred as a consequence, directly or indirectly, of the use and application of any of the contents of this book.

First published: 2011
ISBN: 978-7-117-13973-1/R • 13974

ISBN 978-7-117-13973-1

9 787117 139731 >

Cataloging in Publication Data:
A catalog record for this book is available from the CIP-Database China.

Printed in The People's Republic of China

About the Author

Li Wan-yao

Professor Li Wan-yao graduated from Jiangxi University of TCM in 1982 and obtained her master's degree in acupuncture at Nanjing University of TCM in 1985. In the late 1980's she began research on the effects of acupuncture analgesia at Asahikawa University of Medicine in Japan. Professor Li now serves as Professor of the Guangzhou University of TCM Acupuncture School as well as the Vice Chairperson of a specialized committee on clinical acupuncture for the China Association for Acupuncture and Moxibustion. Some of her other achievements include: Executive Director of a specialized committee on acupotomy, and Vice Chairperson of an apitherapy society of Chinese bee breeders. She has taken a leading role in the clinical and experimental study on the treatment of RA and AS with apitherapy.

Professor Li has been dedicated to acupuncture education and scientific research for many years, and as a prominent educator she instructs the study of channels and collaterals, acupuncture points, acupuncture and moxibustion techniques, and specialized forms of Chinese acupuncture.

She has also participated in the compilation of several works as a Chief Editor, such as: *An Experiential Collection of Clinical Acupuncture in Modern Times*, *Food Therapy and Healthcare in TCM*, *Apitherapy*, and *The Complete Compendium of Acupoint Therapy in Modern Times*. She has also edited other books such as *Geriatrics Treatment with Acupuncture*, *Acupuncture and Tui Na*, and *Introduction to Acupuncture and Moxibustion Techniques*.

About the Author

Li Wan-shan

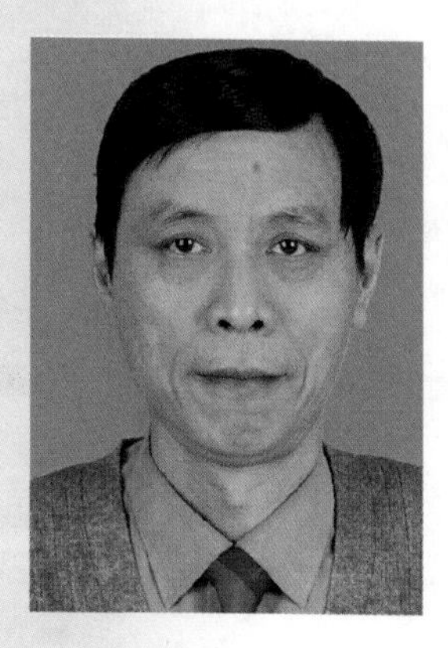

Li Wan-shan is an associate professor at Guangdong Hospital of TCM. After receiving his degree at JiangXi Medical College in 1979, he engaged in pain and anesthesia work at the Nanchang Hospital of Integrative Medicine. Doctor Li later transferred to No.2 Hospital of Guangzhou University of TCM and continued research on the clinical diagnosis and treatment of pain and acupuncture analgesia. He also pursued clinical research on the pain in Japan for two years. Doctor Li's specialized clinical experience in acupuncture analgesia deeply informs his treatment approach to knee osteoarthritis.

Some of his recent publications include: *Observations on the Efficacy of Acupuncture Analgesia*, *Therapeutic Treatments for Osteoarthritis Pain*, *The Role of Acupuncture Anesthesia in Rectal Cancer Surgery*, and *Observations on the Efficacy of Shenmai Injection for Knee Osteoarthiritis*.

Foreword

Acupuncture technique originated in the New Stone Age ten thousand years ago, and it can even be traced back to the Old Stone Age a hundred thousand years ago. The origin of moxibustion can be traced back to the discovery and use of fire 400,000 years ago.

As early as 1500 years ago, in the Northern and Southern Dynasties (the fifth and sixth century A.D.), acupuncture technique and medical books spread to other countries in the world. In the seventeenth century, Chinese acupuncture technique was introduced to Holland, Germany, England, and other Western countries. In the eighteenth and nineteenth century, acupuncture was further propagated in Western countries, as medical doctors in France, Britain, Russia, Italy, Austria, and other countries started to treat diseases, and publish books about acupuncture and moxibustion.

In recent decades, great attention has been paid to the new achievements of Chinese acupuncture and acupuncture anesthesia, especially by medical personnel around the world. Medical organizations from many countries have sent medical experts and scholars to China to learn. A new wave of acupuncture and moxibustion study has emerged like never before. Currently, acupuncture and moxibustion is recognized as effective in treating more than 200 different kinds of diseases, as well as tobacco withdrawal, alcohol withdrawal, drug withdrawal, anti-aging, weight reduction, and cosmetology. Preliminary results have been seen in treating AIDS with acupuncture in the USA and Germany.

In 1975, the World Health Organization asked the Ministry of Health of the People's Republic of China to set up international acupuncture training courses in Beijing, Shanghai, and Nanjing, with English, Japanese, and French as the languages of instruction. This was warmly welcomed by foreign physicians. Currently, international training courses in acupuncture are available in TCM colleges throughout many provinces. China has trained two to three hundred thousand acupuncture practitioners from over 160 countries and regions. After returning to their home countries, many began to treat people with positive results. Academic associations, scientific institutes and schools of acupuncture, and published professional journals began appearing in these countries.

The People's Medical Publishing House (PMPH) saw the need to develop clinical, teaching, and scientific research in acupuncture. The publishing house has determined to launch a series of clinical acupuncture books in foreign countries. These books are chief-edited by Wang Ling-ling, the general-director of the Clinical Branch of the Chinese Association of Acupuncture and Moxibustion (CAAM), and Wang Qi-cai, the chief secretary of the Clinical Branch of CAAM, and compiled by experienced acupuncture experts.

For the first step, thirteen books such as the acupuncture treatment for headache, insomnia, depression, stroke, asthma, knee osteoarthritis, lumbago, trigeminal neuralgia, and obesity are selected by editorial board members from among the most commonly seen diseases in clinical practice. Every book contains nine chapters, consisting of the Chinese medicine and Western medicine approach to the disease, syndrome differentiation and treatment, prognosis, prevention and regulation, clinical experience of other renowned acupuncturists, perspectives of integrative medicine, selected quotes from ancient TCM texts, and modern research. This gives the reader a rather complete understanding of the disease.

The series of books is mainly geared toward the clinical acupuncturist, and can be used as reference books for teaching staff and students in TCM colleges. However, mistakes are sometimes inevitable by individuals following the books. Therefore, only licensed acupuncturists are advised to employ it in clinical practice. Due to laws and regulations in individual countries and regions, some therapies in the books may be limited or even forbidden to use.

We sincerely hope that these books will inspire and help you.

Wang Ling-ling, General-director of the Clinical Branch of CAAM

Wang Qi-cai, Chief-secretary of the Clinical Branch of CAAM

June 1, 2010

Preface

Knee osteoarthritis (Knee OA, KOA), also known as knee hyperosteogeny, knee joint disease or knee degenerative osteoarthropathy are common diseases observed in the clinic of orthopedics and traumatology with high morbidity and disability rates, which is one of the important reasons for affecting the ability to work and quality of life of middle aged and elderly people. The incidence of Knee OA is one of the most common diseases worldwide and is distributed throughout the entire world. With the increase of life expectancy and the change of population structure, the incidence of Knee OA has been increasing and the symptoms have been worsening with age, which strongly influences the quality of life of the patients afflicted with this disease. Therefore, Knee OA has become a global public health concern and the prevention and treatment has become an international concern.

Without a definite name, KOA was ascribed to "*bì* syndrome", "bone *bì*", and "knee pain" in TCM and its symptoms and pathogenesis were described in classic texts as well. *Basic Questions-Great Treatise on the Correspondences and Manifestations of yin and yang* (*Sù Wèn-Yīn Yáng Yīng Xiàng Dà Lùn*, 素问·阴阳应象大论) states: "kidney governs bone which is also supplied by the kidney and generates marrow". *Basic Questions-Treatise on Acupuncture Manipulation* (*Sù Wèn-Cháng Cì Jié Lùn Piān*, 素问·长刺节论篇) also states: "The location of the disease is in the bone which is too heavy to lift and the marrow suffers from aching pain because of the arrival of cold. So it is called bone *bì*". *Comprehensive Medicine According to Master Zhang (Zhāng Shì Yī Tōng,* 张氏医通) states: "the knee is the residence of sinew and knee pain is invariably attributed to deficiency of the liver and kidney. Deficiency conditions always tend to be attacked by wind, cold and dampness".

At the present time, the treatment of Knee OA is a critical issue. In addition to symptomatic treatment, easing the symptoms and operational intervention, there is lack of any effective treatment. However, Traditional Chinese Medicine has its own unique understanding of the etiology and pathomechanism, which produces satisfactory results in the prevention and treatment of the disease by means of syndrome differentiation and comprehensive treatment, treating both internally and externally.

As a result, this book is edited. It has nine chapters in total, which discuss the primary symptoms and pathological mechanisms of Knee OA from a traditional Chinese medicine and modern medicine viewpoint and focuses on the syndrome differentiation of Chinese Medicine and the basic methods of acupuncture as well as other distinctive and effective treatments. The latter portion of the book is accompanied by typical cases observed in the clinic to further support the superiority of TCM in treating this disease. Moreover, it expounds on the prevention and the recuperation of this disease utilizing the principles of health care provided by TCM. The book also expounds upon acupuncture therapeutics, advanced treatments and the acupuncture expertise from some of the finest acupuncture masters.

Afterwards, it is our hope to reduce the difficulty, obstacles and experience of treating Knee OA with the combination of TCM and Western Medicine. With the development of medical science, people need to have a thorough and accurate understanding of this disease. In the end of the book, the newest pathomechanism will have been illustrated in a chapter of modern research. And in this chapter, we also summarize the principles and methods of treating Knee OA with acupuncture of TCM. At last, the target we have aimed at has been presented on the basis of comparison between domestic and overseas research.

During the compilation of this book, we have referred to a variety of writings from many physicians, and extracted information from various reports in medical journals and from the internet. Many of the therapeutic documents of TCM have been collected and catalogued. Here we would like to thank all of the authors that we mentioned above. We have strived to promote complete and accurate information. However, due to limited time and ability, there is the probability of some flaws. It is hoped that all the readers and experts that read this book will kindly point out and rectify the flaws in the book.

Li Wan-yao & Li Wan-shan

December 15, 2010

Table of Contents

Chapter 5

Preventive Healthcare

Chapter 6

Clinical Experience of Renowned Acupuncturists

Chapter 7

Perspectives of Integrative Medicine

Chapter 8

Selected Quotes from Classical Texts

Chapter 9

Modern Research

Chapter 1

The Western Medical Perspective

Brief Description

Knee osteoarthiritis (KOA) is the most common joint disorder occurring in 60% to 90% of individuals older than 65 years of age. Different races appear to have different incidences of knee OA. For example, the Chinese in Hong Kong have a lower incidence of OA than do Caucasians; knee osteoarthritis is also more common among African-American women than Caucasians, and certain occupations that require repeated joint stressors predispose one to early OA. Gender and age are also dominant prevailing risk factors. In a radiographic survey of women less than 45 years old, only 2% had OA; between the age of 45 to 64 years old, however, the prevalence was 3%, and for those older than 65 years of age it reached 68%. Obesity is another powerful risk factor. Researchers have proven that those whose body mass index (BMI) increased by 20% were at higher risk for developing knee OA in the ensuing 36 years at 1.5 for men and 2.1 for women.

Etiology and Pathology

Known as degenerative arthritis, knee OA is characterized by the progressive loss of articular cartilage. It can affect many tissues including the subchondral bone, synovium, meniscus, ligaments and supporting neuromuscular apparatus including the cartilage. Articular cartilage is composed of two macromolecular species: proteoglycans (PGs) and collagen. Interleukin-1 (IL-1) is considered as the driving force of the turnover of normal cartilage. Insulin-like growth factor-1 (IGF-1) and transforming growth factor achieve their functions of regulating, degradation and synthesis through stimulating biosynthesis of PGs, meanwhile, down-regulating chondrocyte receptors for IL-1. Biochemical data has confirmed that the primary changes in knee OA begin by affecting the cartilage. The extracellular matrix of articular cartilage, cytokines and a variety of other biological mediators change, all of which affect chondrocyte metabolism. When the articular cartilage of the knee joint is overloaded, its structure and composition are comprised and apparently change, which easily results in knee OA.

Diagnostics and Treatment

Pain is the characteristic feature of knee OA. Pain is initially aggravated with activity accompanied with deep aching and discomfort. Typically, it is improved with rest and is localized to the involved joint. Nocturnal pain appears as the disease progresses.

Its major features include stiffness, particularly in the morning or after prolonged inactivity, but usually last less than 20 minutes without systemic manifestations. Joint pain can be caused by many reasons, some of which follow: inflammation of the synovium; microfractures of the subchondral bone; formation of osteophytes, stretching of the ligaments, distention of the articular capsule of the knee joint, and muscle spasms.

When performing a physical examination of the knee joint affected by OA, it may reveal localized tenderness and bony or soft tissue swelling. Bony crepitation on motion and range of motion limitation of the joint are additional characteristic features. In the advanced stage of OA, there may be gross physical deformity, bony hypertrophy and complete loss of joint motion.

The diagnosis of knee OA is usually based on history, physical examination and characteristic radiographic analysis. The major particular features include narrowing of the joint space, subchondral bone sclerosis and osteophytosis. There is often great disparity between the severity of radiographic findings, the severity of symptoms, and the functional ability of the knee diagnosed with OA.

It is usually not difficult to differentiate knee OA from a systemic rheumatic disease such as rheumatoid arthritis, because, in the latter disease, joint involvement is usually symmetric and poly-articular, and there are also constitutional features such as prolonged morning stiffness, fatigue, weight loss and/or fever.

Management of this condition includes non-pharmacologic measures and drug therapy, which should be determined according to the condition of the individual patient's disease progression.

Non-Pharmacologic Therapy

For patients with mild disease, education and physical measures are of benefit. Education includes mental reassurance and advice on joint protection and exercise. Patients can participate safely in conditioning exercises to improve fitness and health without increasing their joint pain, as strengthening of periarticular muscles can protect the articular cartilage from stress. Reduction of joint loading includes correction of poor posture, avoidance of excessive loading, prolonged standing, kneeling and squatting. Weight loss may retard knee OA progression in obese patients. The use of assistive devices such as a cane can also provide important joint protection. Occasionally, analgesics may be prescribed as required.

Pharmacologic Therapy

Pharmacologic therapy for knee OA provides symptomatic relief but has not been shown to successfully alter the course of disease. Available pharmacologic therapies include simple analgesics and non-steroidal anti-inflammatory drugs (NSAIDs). Simple analgesics are useful and generally well tolerated in moderate cases of knee OA. The use of NSAIDs has been controversial, largely because of the debate over their risks and benefits. The side effects of these agents such as gastrointestinal symptoms, ulcerations, hemorrhage and death can not be neglected, especially when recommending them to older populations. Intra-articular injections of glucocorticoid or synthetic hyaluronic acid appear to provide modest symptomatic benefit.

Generally, non-pharmacologic management is the foundation of treatment for OA and is as important as, and often more important than drug treatment.

Disadvantages of Western Medical Treatment

Over the years, pharmacologic therapy for knee OA has been treated with NSAIDs. The mechanism of NSAIDs is fulfilled by inhibiting the synthesization of central PG (Proteoglycans) and tapering off the sensitization of PG to inflammatory mediators such as bradykinin. NSAIDs have more obvious unpleasant effects, such as Adult Respiratory Distress Syndrome (ADRs) and other less desirable side effects to the older population such as gastrointestinal hemorrhage, the most common complication. There is also evidence that after taking salicylic acid agent orally, the synthesization of proteoglycans and hyaluronic acid decreases and the surface of the arthrodial cartilage fiberizes, aggravating the pathological injury of OA. Much research has shown that amino glucose cartilage-protecting drugs could relieve the symptoms of KOA and delay the degradation of arthrodial cartilage simultaneously. The experiments indicate that amino glucose possibly has the dual-action of protecting arthrodial cartilage and providing anti-inflammatory results. Amino glucose is often viewed as the one promising drug among the small amount of inexpensive Western drugs which aim at the pathogenesis of KOA.

The traditional surgical therapy includes Joint debridement, synovectomy, osteotomy and arthrodesis. If the symptoms conform well to the indication, surgical therapies could obtain certain beneficial effects. But all of these procedures fail to reverse the pathologic changes of the knee or reconstruct the injured joint surface. In severe cases surgical therapies have limited effects, or none at all in some cases.

In recent years joint replacement therapy has been considered as a routine operation. In severe cases the treatment consists of surgical reconstruction of the knee or knee replacement. The problem with joint prosthesis is that they wear out over time so it has to be replaced occasionally. Because the prosthesis wears out over time, future operations may be needed and this raises the degree of operational difficulty. For this reason such treatment is often only applied in those people over 60 years of age.

Searching for an appropriate therapy is still a hot point in the study of KOA. In the long-term quest for alternative therapies, Traditional Chinese Medicine has summarized a great number of therapies and formulas, especially for the patient in the early and middle stages. It is a worthwhile effort for clinical workers to focus on the favorable effects of Traditional Chinese Medicine.

Chapter 2

Chinese Medical Perspective

Brief Description

Without a definite name, KOA was ascribed to *bì* (痹) syndrome in TCM and its symptoms and pathogenesis were described in classical texts as well. Its major manifestations are pain, numbness, heaviness, heat sensation, limited range of motion of the muscles and tendons, and occasionally swelling and deformity of the joints. *Bì* syndrome is always caused by channel obstruction and inhibited flow of qi and blood due to insufficient healthy qi, insecurity of the exterior *wei qi* (卫气), and external contraction of wind, cold, dampness and heat. Some manifestations pertain to the categories of "wind-dampness" or "body pain", others could be classified as one disease, such as pain of multiple joints, white tiger bite-like (白虎历节病), and bone *bì*.

Basic Questions - Discussion on bì (*Sù Wèn – Bì Lùn Piān*, 素问•痹论篇) states: "the combination of wind, cold and dampness attacks and results in *bì*". *Basic Questions - Treatise on Acupuncture Manipulation* also states: "The location of the disease is in the bone which is too heavy to lift and the marrow suffers from aching pain caused by the arrival of cold. This is why it is called "bone *bì*".

Essentials from the Golden Cabinet (*Jīn Guì Yào Lüè*, 金匮要略) by Han Dynasty physician Zhang Zhong-jing calls this "pain of multiple joints" (历节).

Etiology and Pathology

In TCM the etiology and pathomechanism of KOA is always understood from the viewpoints of deficiency, a complex combination of deficiency and excess, or root-deficiency with branch-excess. The characteristics of deficiency include multiple insufficiencies of qi, blood, yin and yang of the *zang-fu* organs. On the contrary, the characteristics of excess include: qi stagnation, phlegm coagulation, blood stasis, and external contraction of the six pathogenic factors.

Liver, Kidney and Spleen Depletion as the Foundation of Sinew and Bone Malnourishment

TCM holds the belief that the knee joint is the major confluence of sinews, tendons and muscles and is connected by the channels of the kidney, spleen and liver.The liver stores the blood and governs the tendons. The kidney stores the essence and governs the bones, and the spleen governs transformation, transportation and dominates the muscles.

KOA almost always attacks people who are over middle age. When the male is over 56 years old and the female over 49 years old, the liver and kidney gradually decline and become deficient, causing the sinews and bones to become slack and the tendons to become malnourished, thus failing to soften the joints. The bone marrow then becomes malnourished and fail to strengthen the body. Additionally, the muscles lack domination by the spleen resulting in muscle weakness. Therefore, it is quite clear that the onset of KOA has a close relationship with the weakened condition of the liver, spleen and kidney.

Kidney Deficiency

TCM principles state that the kidney and bones have a close relationship. *The Yellow Emperor's Inner Classic: Basic Questions* (*Huáng Dì Nèi Jīng Sù Wèn*, 黄帝内经素问) states: "The kidney governs the bones and the bones are supplemented by the kidney, this combined interraction produces marrow". Abundant kidney essence ensures proper generation and transformation of marrow thus promoting firm and powerful bone development.

Basic Questions - Discussion on Phlegm (*Sù Wèn - Tán Lùn*, 素问•痰论) points out: "the kidney is a water viscus; when the water is inferior to fire, the bone will become exhausted and the marrow weakened, so the feet are unable to support the body, causing bone-phlegm to occur".

As a result, deficient kidney essence will result in insufficient marrow transformation and bone malnourishment. This condition easily leads to fractures and osteopathia resulting in the vulnerable structure of the bone. KOA is always accompanied by Osteoporosis, and pertains to the category of bone-*bì* in TCM. This provides us evidence that KOA is related to kidney deficiency.

The Yellow Emperor's Inner Classic states: "renal failure results in body exhaustion". This informs us that the physiological decline caused by the deficient kidney could make the tendons unhealthy, transformation of marrow insufficient, articular cartilage and subchondral bone malnourished, and cartilage degenerated, eventually leading to KOA.

Liver Yin and Blood Deficiency

TCM theory states that the liver is related to sinews in the body. The liver takes charge of permeating the essential qi, transported by the spleen, to the sinew and fascia, and makes the joints move smoothly, forcefully and freely. The first tissues affected by this pathological condition (KOA) are the articular cartilage, synovium and subchondral bone. In TCM, All of them belong to the category of "sinew". Thus, KOA is a sinew disease, closely related to the liver according to the theory that states: "the liver controls sinews", and "the knee joint is the residence of sinew".

The relationship between the liver and the sinew were described in classics, such as *Basic Questions - Treatise on the Six Periods and Visceral Manifestation* (*Sù Wèn – Liù Jié Zàng Xiàng Lùn*, 素问•六节藏象论) which states: "the liver has its manifestation in the sinew", and *Basic Questions – A Special Discussion on Channels* (*Sù Wèn – Jīng Mài Bié Lùn*, 素问•经脉别论) states: "after the food has entered the stomach, the nutrients are dispersed to the liver and permeate the sinew". This theory proves that the normal function of sinew depends on the liver's normal function of storing blood and the flowing of qi.

A Handbook on Famous Doctors (*Míng Yī Zhǐ Zhǎng*, 名医指掌) by Ming Dynasty physician Huang-fu Zhong states: "overstraining of the liver corresponds to sinew-consumption", and *Comprehensive Medicine According to Master Zhang* (*Zhāng Shì Yī Tōng*, 张氏医通) states: "The knee is the residence of sinew".

Knee pain is invariably ascribed to deficiency of the liver and kidney. Deficiency conditions always tend to be vulnerable to attack by wind, cold and dampness.

A deficient condition of liver yin and blood always results in malnutrition of the sinews. If the flow of liver qi is excessive, it will restrict earth and the spleen will fail to transform and transport, resulting in local channel blockage by pathogenic damp-turbidity. Meanwhile, yang qi is also blocked, causing a series of symptoms, such as pain, swelling, limb vibration and numbness, which eventually causes inhibited bending and stretching.

Spleen-Stomach Weakness

The spleen and stomach are the foundation of acquired (postnatal) constitution and the spleen dominates the muscles and limbs. The movement of the four limbs and joints relies on the condition of the spleen and stomach. It is generally acknowledged that KOA is caused by transformational hypo-function of the spleen and stomach due to kidney yang depletion and cold-damp encumbering the spleen. After a long illness, this causes retention of internal water-dampness gathering and producing phlegm. The spleen is viewed as the source of production and transformation of qi and blood. Qi is the commander of blood and spleen deficiency leads to qi deficiency, so blood stasis is always caused by weakness of blood flow due to qi deficiency. The mutual cause between phlegm-dampness and blood stasis promotes the development and vicious cycle of KOA.

Qi Stagnation, Blood Stasis and Phlegm Coagulation as Influential Factors in the Process of Morbidity

Qi is the "commander" of blood, and blood is the "mother" of qi. Qi stagnation results in blood stasis, and blood stasis aggravates qi stagnation.

Basic Questions - Great Treatise on the Correspondences and Manifestations of yin and yang (*Sù Wèn – Yīn Yáng Yīng Xiàng Dà Lùn*, 素问•阴阳应象大论) states: "qi damage leads to pain and constitutional damage produces swelling".

Vessel damage caused by chronic overstrain and repeated stimulation of external injury makes the blood flow outside the vessels and blocks the channels, which leads to qi stagnation and blood stasis. Inhibited flow of qi caused by external attack of pathogens and channel blockage makes the clear fluid gather and produce phlegm-damp which transforms into phlegm stasis. Phlegm stasis is then retained at the joints and bones which aggravates the obstruction, making the pain become worse.

It is written in *Basic Questions - Discussion on bì* that: "Obstructions retained between the tendons and bones are responsible for long-term pain".

The onset and development of KOA can also be casued by degeneration of articular cartilage due to depletion of the liver, spleen and kidney, inhibited flow of qi and blood, and phlegm coagulation in the channels resulting in malnutrition of the knee joint and its peripheral tissues.

External Wind-Cold and Dampness Obstructing the Channels

KOA pertains to the category of bone *bì* and knee pain. *Treatise on Diseases, Patterns, and Formulas Related to the Unification of the Three Etiologies* (*Sān Yīn Jí Yī Bìng Zhèng Fāng Lùn*, 三因极一病证方论) states: "the three external pathogens invade the channels", and "the bone's response to this is that they become too heavy to be lifted; the vessel's response is congealing of blood failing to flow; the sinew's response is flexion instead of extension and the muscle's response is numbness".

As a result, after middle age KOA always results from the attack of wind-cold-dampness when the condition of depletion occurs in the liver, kidney, qi and blood. Clinically, this condition is seen frequently in people whose vessels are easily coagulated and channels obstructed by collective invasion of cold-dampness and deficiency of interstitial spaces. After prolonged exposure to cold-damp environments and overstrain, it finally results in the sensation of heavy pain which is difficult to recover from.

Overall, after a long history of study and exploration by our ancestors, we can conclude that KOA is closely related to deficiency, external pathogenic factors, and stasis. Deficiency conditions of the liver, spleen and kidney are the fundamental cause of the external factors of wind-cold and dampness, where blood stasis is the pathological outcome during the course of the disease. Pathogenic factors and stasis can lead to deficiency, while the condition of deficiency interferes with the elimination of pathogenic factors and stasis. As stated above, it is easy to see that pathogenic factors and stasis can mutually affect one another.

Differential Diagnosis

Wěi (痿) syndrome is a disorder marked by weakness of the limbs and tendons, numbness, atrophied and flaccid limbs, difficult unsmooth movement, even paralysis but without pain. Differing in nature, *bì* syndrome is marked by obvious pain all the time with numb and atrophied limbs manifesting after long term illness. The primary point to distinguish the difference between the two syndromes is the sensation of pain. The second point, whether the muscular atrophy is severe or not. *Bì* syndrome is characterized by pain, while *wěi* syndrome is marked by weakness and flaccidity. Their etiologies, pathogenesis and treatment are also different, and treatment is based on the manifestation and differentiation of symptoms.

Chapter 3

Syndrome Differentiation and Treatment

Knee *bì* (痹) should be treated in accordance with its characteristic pattern of "root deficiency with branch blockage". In the remission stage, the liver and kidney are usually insufficient, and the channels are blocked by phlegm and stasis. In the acute stage, knee *bì* is usually affected by wind, cold, and dampness, occasionally manifesting with signs and symptoms of damp heat pouring downward.

The wind-cold-damp *bì* syndrome can be subdivided into three distinct types: wandering (wind) *bì*, painful (cold) *bì* and fixed (damp) *bì* all of which are independent of the prevailing pathogenic factor. Once the treatment addresses the main pattern, excellent effects can be obtained.

Syndrome Differentiation

Light and mild knee pain which is accompanied by sensations of soreness and heaviness can become aggravated under the burden of a heavy load. The knee joint will then become stiff and swollen, and will not bend or straighten freely.

Wind-Cold-Damp Bì

1. Migratory (Wandering) bì:

This pattern exhibits inhibited flexion and extension of the knee joint, painful and sore muscles, and it usually affects multiple joints simultaneously. The migrating pain is caused by the dominant nature of wind. Symptoms include an aversion to wind, fever, a pale tongue with a thin white coating, and a floating moderate pulse.

2. Painful (Cold) bì:

This pattern is characterized by severe pain which remains localized at a fixed location and is aggravated by cold and alleviated by warmth. Movement is inhibited without localized redness of the skin or heat signs. Signs and symptoms include a pale red tongue with a white coating, and a wiry and tight or deep pulse.

3. Fixed (Damp) bì:

This pattern manifests with fixed, localized soreness and pain with a heavy sensation, and swelling or numbness of the knee joint and limited range of motion caused by the predominance of dampness. The pattern is aggravated on rainy and cloudy days, often resulting in recurrence. The tongue will be pale red with a greasy thin white coating; the pulse is soggy and moderate.

Heat *Bì*

Heat *bì* manifests with arthralgia, inhibited range of motion, and erythema manifesting as localized redness and heat of the skin caused by dilation and congestion of the capillaries, often coexisting with subcutaneous nodules. These symptoms are also often accompanied by fever, aversion to wind, sweating, thirst, and restlessness. The tongue will appear red with a greasy yellow coating; the pulse is rapid and floating or slippery.

Phlegm Stasis and Obstruction

This pattern is observed in prolonged *bì* syndrome, with a fixed stabbing pain, dull purple-colored swelling, numbness and the sensation of hardness and heaviness on the surface of the joint, or a stiff and deformed joint. There will be inhibited movement with sclerosis and ecchymosis. Other signs and symptoms include a blackish complexion, eyelid edema and/or chest oppression with profuse phlegm. The tongue will appear dull purple with ecchymosis and a greasy white coating; the pulse is wiry and choppy.

Liver and Kidney Deficiency

This pattern manifests with prolonged impairment exhibiting inhibited movement, lean and wasting musculature, lumbar weakness, and flaccid knees. Other signs and symptoms include aversion to cold with cold limbs, impotence, seminal emission, and steaming bone fever with vexation with a dry mouth. The tongue will be pale and red with a thin white or moist coating; the pulse is deep and thready or weak and rapid.

Cold Congealing due to Yang Deficiency

This pattern manifests with knee pain which is aggravated by cold and alleviated by warmth, a bloated and pale complexion, a preference for warmth and an aversion to cold. Other signs and symptoms include a lack of strength with no desire to eat, and diarrhea. The tongue will be pale with a white moist coating; the pulse is deep, thready and weak.

Treatment

Acupuncture and Moxibustion

1. Standard Acupuncture Treatment

【Treatment Principles】

The general treatment principle is to unblock the channels and quicken blood to relieve pain.

Migratory *bì*: invigorate blood to dispel wind.

Painful *bì*: warm the channels and dissipate cold.

Fixed *bì*: eliminate dampness and remove turbidity.

Heat *bì*: clear heat and relieve swelling.

Phlegm stasis and obstruction: invigorate blood and remove phlegm.

Liver and kidney deficiency: supplement and boost the liver and kidney.

Cold congealing due to yang deficiency: warm yang and dissipate cold.

Both acupuncture and moxibustion using the drainage method can be applied for patterns of migratory *bì*, painful *bì* and fixed *bì*. Conversely, in heat *bì*, acupuncture with the drainage method is used without moxibustion. Acupuncture and moxibustion with supplementing or the draining methods can be used for patterns of phlegm stasis and obstruction. Additionally, the pattern of cold congealing due to yang deficiency should be treated with moxibustion using supplementation.

【Basic Point Prescription】

ST 34 (*liáng qiū*)	SP 10 (*xuè hăi*)	*xī yăn* (EX-LE5, 膝眼)
ashi points		

【Explanation】

ST 34 (*liáng qiū*) is the *xi*-cleft point of the foot *yangming* stomach channel where the channel qi deeply converges and passes through the knee area, so diseases of the knee area can be treated with ST 34 (*liáng qiū*) by means of regulating the channel qi of the stomach.

The spleen controls the blood and the *taiyin* channel and has an abundant amount of blood with lesser qi which has a mutual interior-exterior relationship with the *yangming* channel, which is the channel with an abundant amount of blood and qi. So as the acupuncture point of the spleen channel, SP 10 (*xuè hăi*) is effective in treating diseases related to the blood level by invigorating the blood, dispelling wind, moving qi and unblocking the collaterals to relieve pain.

The occurrence of *bì* syndrome is often caused by the collective invasion of wind-cold-damp which leads to qi stagnation and blood stasis, so the principle of point selection is based on the theory that "wind will naturally be eliminated when the blood circulates smoothly".

The two *xī yăn* (膝眼) acupoints located on the lower border of the patella and on the medial and lateral side of the patellar ligament can be used to treat diseases of the knee joint, while the lateral *xī yăn* point is actually ST 35 (*dú bí*), which can regulate *yangming* channel qi. Knee *bì* relates to sinew *bì* whose principle of treatment has been discussed in *The Yellow Emperor's Inner Classic*, which states: "determine the points according to the tender spots". *Ashi* points are frequently used in clinical practice.

【Point Modifications】

- Add BL 17(*gé shù*) for migratory *bì* to invigorate and regulate the blood.
- Add BL 23 (*shèn shù*) for painful *bì* to warm and supplement yang qi, remove cold and supplement the kidney.
- Add ST 36 (*zú sān lǐ*) and SP 9 (*yīn líng quán*) for fixed *bì* to fortify the spleen and eliminate dampness.
- Add LI 11 (*qū chí*) and DU 14 (*dà zhuī*) to drain heat.
- Add ST 40 (*fēng lóng*), and BL 20 (*pí shù*) for phlegm stasis and obstruction.
- Add BL 18 (*gān shù*), BL 23 (*shèn shù*), KI 3 (*tài xī*) and LV 3 (*tài chōng*) for liver and kidney deficiency.
- Add RN 4 (*guān yuán*), RN 6 (*qì hǎi*) and DU 4 (*mìng mén*) for cold congealing due to yang deficiency.

【Manipulations】

Generally, the knee joint should be bent at an angle when receiving acupuncture. When the patient is lying down, a square cushion or something of the sort must be laid under the knee joint in order to help the joint to flex naturally. ST 34 (*liáng qiū*) and SP 10 (*xuè hǎi*) should be needled perpendicularly to 1-2 *cùn* with lifting, thrusting and twirling manipulations to induce drainage; this should produce localized soreness and distention or a needle sensation spreading around the knee joint and cavity.

Insert *xī yǎn* (膝眼) 1-2 *cùn* obliquely from the anterior lateral side of the knee to the posterior medial side of the knee and manipulate with the even method. The needling sensation should spread to the whole knee joint, or radiate downward.

Ashi points can be inserted using the approach of short needling, which is inserting the needle from a superficial level to the deep level while simultaneously shaking the needle handle until the needle arrives at the bone, then needling up and down near the periosteum as if rubbing and scraping the bone.

Other additional points can be needled using standard routine methods. Electro-acupuncture can be added to the treatment when flexion and extension are inhibited. The principal points are always used as the negative and positive poles and should be stimulated with an intermittent pulse wave setting.

2. Traditional Moxibustion Therapy

Moxibustion is a therapy that utilizes cauterization or heating on certain areas of the body. It regulates the functions of the channels, thus it is used to treat and prevent disease. *Precepts for Physicians* (*Yī Mén Fǎ Lǜ*, 医门法律) states: "Moxibustion must be used when medicinals and acupuncture have no effect."

There are many methods of moxibustion, including cone moxibustion, mild moxibustion, stick moxibustion, warming-needle moxibustion, moxibustion with a moxa burner and so on. As for the pattern of wind-cold-damp *bì*, use the acupoints which can warm the channels, unblock the vessels, dissipate cold, eliminate dampness, invigorate the blood and relieve pain. According to the theory of TCM, for the pattern of heat *bì* we should adhere to the principle that "fire constraint must be dispersed".

【Basic Point Prescription】

The following acupoints located on the foot *yangming* stomach channel, foot *taiyin* spleen channel, foot *shaoyang* gallbladder channel and other local points are frequently used:

BL 40 (*wěi zhōng*)	ST 34 (*liáng qiū*)	*xī yǎn* (EX-LE5)
GB 34 (*yáng líng quán*)	SP 9 (*yīn líng quán*)	*hè dǐng* (EX-LE2)
SP 10 (*xuè hǎi*)	ST 36 (*zú sān lǐ*)	ST 35 (*dú bí*)
GB 33 (*xī yáng guān*)	*ashi* points	

【Indications】

Patterns of wind-cold-damp *bì*, phlegm stasis and obstruction, liver and kidney deficiency, and cold congealing due to yang deficiency.

【Manipulations】

Method Ⅰ: Warming Needle Moxibustion

Selected points: *nèi xī yǎn* (内膝眼), ST 35 (*dú bí*), *ashi* points, ST 34 (*liáng qiū*), SP 10 (*xuè hǎi*), GB 34 (*yáng líng quán*), SP 9 (*yīn líng quán*), and ST 36 (*zú sān lǐ*).

The patient is sitting or lying in a supine position with the knee joint flexed naturally. Disinfect the points to be treated and insert the needle rapidly using the twirling method. After the qi arrives, manipulate the needle using the even method until the patient senses a slight soreness and distention after which the needles are retained for 30 minutes. During the needle retention phase, affix a 2 cm long section of moxa to the handle of the needle. Withdraw the needle after 2 applications of moxa have burned out completely and the ash is removed. Simultaneous acupuncture and moxibustion is an especially effective technique for this condition.

Method Ⅱ: Indirect Moxibustion

Indirect moxibustion is a technique that is often applied to the knee area. Prepare some pounded or sliced ginger and form some moxa wool into the desired number of cones before treatment begins. The patient exposes their knee area and the practitioner lays the ginger onto the affected area, or onto the selected acupoints. Then moxa cones are placed on the ginger and ignited. When the moxa cone extinguishes, the cone is replaced and the process continues until the number of cones prescribed have completely burned out. The skin turns a reddish color, without blister formation. Besides the ginger, a *fù zǐ* cake can also be used for moxibustion.

Method Ⅲ: Cone Moxibustion

The selected points are *ashi* points, ST 35 (*dú bí*), *xī yǎn* (膝眼), ST 34(*liáng qiū*), SP 10 (*xuè hǎi*), and ST 36 (*zú sān lǐ*).

Shape some moxa wool into a cone with the bottom diameter of 1-2 cm. Then put the cone on the point that is to be treated and ignite it. Let the moxa burn down until the local skin becomes reddish in color and the patient feels a warming sensation with no sensation of burning.

Since moxibustion is a simple and convenient therapy, it can be used alone or in combination with acupuncture, electro-acupuncture, knife-needling, and Chinese medicinals.

Related Acupuncture Therapies

1. Modern Heat-Sensitization Moxibustion Therapy

Heat sensitization is a popular therapy which refers to acupoints with the unusual sensation of non-localized and non-superficial sensation, or even no heat sensation at all when moxa heat is applied. The responses of a heat-sensitized acupoint to heat are heat penetration, heat diffusion and heat transmission, or no heat sensation at all.

One survey has shown that the occurrence rate of heat sensitization on an acupoint occurs in about 70% of the cases with KOA. As a pathological response, the results of heat sensatization are not coincidental with the acupoint, but can be located by referring to the coordinated system of acupoints.

【Prescription】

Acupoints of the whole body, localized heat treatment on sensitized points.

【Indications】

Patterns of wind-cold-damp *bì*, phlegm stasis and obstruction, liver and kidney deficiency, and cold congealing due to yang deficiency.

【Procedure】

First, the doctor should find the heat sensitized acupoints before applying moxibustion. Then treat the tender acupoints adjacent to the knee joint with subcutaneous nodules as the center; apply mild moxibustion with a moxa stick. When the patient feels the moxa heat penetrate deeply into the skin, the point is then at the heat sensitization point. Repeat the steps above until all of the heat sensitized points are treated. After applying mild moxibustion to each heat sensitized point the phenomenon of heat penetration disappears. The time for one dose of moxibustion varies from 5 to 100 minutes. The standard is dictated by the reaction of the heat-sensitization of the acupoint. This therapy is usually applied once a day.

In addition to moving qi, invigorating blood, dispersing stasis and dissipating masses, modern research has confirmed that heat sensitizing moxibustion can improve local micro-circulation and relieve inflammation. Furthermore, it can relax muscle spasticity, enhance dynamic stability and improve symptoms. Because of its efficacy, heat sensitization moxibustion is superior to many other routine therapies.

2. Cupping Therapy

Cupping therapy is also known as the cup-sucking method, fire cupping or in ancient times, animal horn cupping. Cupping is a therapy in which a jar is attached to the skin surface using the negative pressure created by introducing a flame into the cup or by some other means of suction so as to form localized congestion or blood stagnation. Specifically, it can be divided into three methods:

I. Retention Cupping

II. Bleeding Cupping

III. Medicated Cupping

【Prescription】

From the perspective of the theory of channels "the indication is where the acupoint is located", these acupoints can relax the sinews, quicken and unblock the collaterals, and relieve pain around the knee joints.

For example, SP 10 (*xuè hǎi*), GB 34 (*yáng líng quán*), SP 9 (*yīn líng quán*), ST 36 (*zú sān lǐ*) and other points around the affected area are often used.

【Indications】

Used for the patterns of wind-cold-damp *bì*, liver and kidney deficiency, and cold congealing due to yang deficiency. Bleeding cupping is applicable to heat *bì* and patterns of phlegm stasis with obstruction.

【Manipulations】

Method I: Retention Cupping

Attach the cup to the knee joint area and the back along the *du mai* and retain for approximarely 10 minutes.

Method II: Bleeding Cupping

Start by pricking *ashi* points around the knee joint to induce bleeding, and then attach the cup to the points.

Method III: Medicated Cupping

Medicated cupping has gradually developed from cupping therapy and is widely used in clinical practice today. Although it differs from the usual external application of Chinese medicine, ion-introduction therapy, fumigation, washing, and medicinal cupping, it generally pertains to external therapy which complies with treatment in accordance with pattern differentiation.

Wu Shang-Xian in the Qing Dynasty stated that this is the "same theory and nature, but a miraculous therapeutic method", and "like internal therapy, external therapy is intended to to be the primary foundation. What is this foundation on earth? Attention is focused on yin-yang and *zang-fu*", and also, "the theory of external therapy is the same as internal therapy; the medicinals used in external therapy are the same as the medicinals used internally. The only difference is the treatment approach".

That is to say, external therapy and internal therapy are both in compliance with the same etiologies, pathomechanisms and medicinal formulas, but they do have different approaches regarding the administration of the medicinals and their absorption.

【Medicinals】

Medicinals must be chosen for treatment according to pattern differentiation.

Basic Medicinals

川牛膝	*chuān niú xī*	Radix Cyathulae
炒穿山甲	*chuān shān jiă*	Squama Manitis

络石藤	*luò shí téng*	Caulis Trachelospermi
土茯苓	*tŭ fú líng*	Rhizoma Smilacis Glabrae
人参	*rén shēn*	Radix et Rhizoma Ginseng
川芎	*chuān xiōng*	Rhizoma Chuanxiong
乳香	*rŭ xiāng*	Olibanum, frankincense
没药	*mò yào*	Myrrha, myrrh
天南星	*tiān nán xīng*	Rhizoma Arisaematis
五加皮	*wŭ jiā pí*	Cortex Acanthopanacis
三七	*sān qī*	Herba Rhodiolae Henryi
冰片	*bīng piàn*	Borneolum Syntheticum

➢ Add *qiāng huó* (Rhizoma et Radix Notopterygii), *fáng fēng* (Radix Saposhnikoviae), *dú huó* (Radix Angelicae Pubescentis), *qín jiāo* (Radix Gentianae Macrophyllae), *zhì chuān wū* (Radix Aconiti Praeparata), *căo wū* (Radix Aconiti Kusnezoffii), *xì xīn* (Radix et Rhizoma Asari), and *fù zĭ* (Radix Aconiti Lateralis Praeparata) for wind-cold-damp *bì*.

➢ Add *xiān máo* (Rhizoma Curculiginis), *xiān líng pí* (Herba Epimedii), *mă qián zĭ* (Semen Strychni), *fù zĭ* (Radix Aconiti Lateralis Praeparata), *hàn lián căo* (Herba Ecliptae), *nǚ zhēn zĭ* (Fructus Ligustri Lucidi), *hé shŏu wū* (Radix Polygoni Multiflori), *sāng jì shēng* (Herba Taxilli), and *ròu cōng róng* (Herba Cistanches) for liver and kidney deficiency.

➢ Add *chuān wū* (Radix Aconiti), and *wú gōng* (Scolopendra) for extreme pain.

➢ Add *dān shēn* (Radix et Rhizoma Salviae Miltiorrhizae) for severe blood stasis.

All medicinals are ground, crushed and sifted out for reserve. Medicated cupping is the combination of using medicinals and cupping together, and there are two methods described as follows:

The first method is the cup boiling method in conjunction with medicinals.

It is also called "medicated bamboo cupping". Using a bamboo cup or a suction cup (glass cups being most commonly used in modern clinical practice), place the medicinal ingredients into a bag or container and place the pouch and the bamboo cup into boiling water for 10-15 minutes. This practice is known as water cupping therapy. Take the bamboo cup out with a pair of tweezers, pour the excess liquid out, cover the opening with a folded towel, and then place the cup rapidly onto the skin.

The second method is to place an appropriate amount of medicinal liquid into the suction cup (about 1/2 to 2/3 of the cups capacity) and to place the cup at the position that needs treatment. Then draw the air out and attach the cup to the skin; or, put the medicinal liquid into the glass cup (about 1/3-1/2 of the capacity) and employ the flash fire cupping method to attach the cup to the skin.

Another method is to put 3-5 drops of medicinal wine into the cup and rotate the cup in order to make the medicinal wine adhere to the upper area of the internal surface of the cup; after that ignite the wine and quickly place the cup onto the skin. The medicinal liquid used commonly in clinical practice includes pepper water, tincture of Radix Zanthoxyli, fresh ginger juice, or an anti-wind-damp medicinal wine.

Basic Questions - Discussion on Cutaneous Regions (*Sù Wèn – Pí Bù Lùn*, 素问·皮部论) states: "The collateral vessels of the twelve channels pertain to the cutaneous regions individually. All diseases start from the skin and body hair. Pathogenic invasion causes the interstitial spaces to open which allows the pathogen to invade the collaterals which are then retained internally. Then the pathogens transmit to the channels, are retained and fail to be removed which leads to transmission to the *zang* and affect both the the stomach and the intestines".

In turn, the diseases of the *zang-fu* organs affect certain areas of the body surface. So it is not difficult to conclude that both the functional activities and the exterior-interior relationship of the *zang-fu* organs are founded on the connection of the channels. Medicinal cupping makes use of its unique efficacy through treatment of the channels.

Medicinal cupping therapy combines acupoint selection with cupping and takes advantage of the negative pressure applied to the skin of the treated area by expanding the surface area of the skin in order to reinforce the medicinal effect. Negative pressure, expanded capillaries, and localized congestion increases medicinal osmosis by means of the warming and pressurization effect inside the cup. The medicinal effect can stimulate the acupoint and channel due to its acrid property and dissipative nature in order to relieve pain, invigorate blood, dissolve stasis, secure the foundation, and reinforce healthy qi.

Modern studies also explain that the negative pressure created by cupping acts to facilitate congestion and bursting of capillaries, which produces hemostasis. The dissolution and absorption of the ruptured erythrocytes brings about a series of optimal results as the body introduces histamine into the blood, which then flows throughout the whole body. The relevant viscera and organs will be stimulated, and their physiological function will be reinforced as well. Furthermore, the mechanical stimulation caused by cupping is transmitted to the central nervous system by afferent nerves causing the brain to regulate excitement and inhibition, and balance functional activities of the *zang-fu* organs. These effects promote the recovery of normal bodily function while also reinforcing the immune system.

3. Jin's 3-Needle Acupuncture Point Therapy

The 3-needle method was created by renowned acupuncturist Professor Jin Rui. The 3-needle method was developed through repeated and systematic research and by collecting the essential experience of many great acupuncturists from throughout history. Dr. Jin also summarized clinical research findings from both domestic and foreign practitioners. Although his technique is renowned as a flourishing new Cantonese school of acupuncture therapy, Jin's 3-needle therapy is actually a traditional acupuncture technique. Jin's 3-needle therapy is named for taking 3 acupuncture points as one group in a prescription, which is different from the traditional types of point combinations. One group of Jin's 3-needles is mainly used to treat knee diseases which pertain to *bì* syndrome such as KOA, knee strain, RA, and sprain of the knee joint.

In terms of *bì* syndrome, *The Yellow Emperor's Inner Classic* (*Huáng Dì Nèi Jīng*, 黄帝内经) stated that the principle of "determine the tender points, determine frequency of occurance, then insert and withdraw the fire needle rapidly"; this indicates the principle for point selection. "*Bì* syndrome above the loins leads to the hand *taiyin* and *yangming* channels" and "*Bì* syndrome below the loins resorts to the foot *taiyin* and *yangming* channels". Local points can regulate the channel qi, disperse pathogenic qi, reduce swelling and relieve pain, but they must be selected based on the affected area and the corresponding channel.

【Prescription】

The knee's 3-needle technique is composed of *nèi xī yăn* (内膝眼), ST 35 (*dú bí*), ST 34 (*liáng qiū*)), and SP 10 (*xuè hăi*).

【Indications】

The same as for standard needling.

【Manipulations】

Flex the knee when performing the knee 3-needle technique. Place a high pillow below the knee joint with the patient in a supine position in order to help the knee flex naturally.

First, insert *nèi xī yăn* (内膝眼), ST 35 (*dú bí*), ST 34 (*liáng qiū*), and then SP 10 (*xuè hăi*) and *xī yăn* (膝眼); each can be inserted approximately1-1.5 *cùn* inward, but do not insert the needle into the articular cavity. ST 34 (*liáng qiū*), and SP 10 (*xuè hăi*) can be needled perpendicularly to about 1.2 *cùn* until qi is obtained.

Bì syndrome is often divided into wind *bì*, painful *bì*, fixed *bì* and heat *bì* according to the predominant etiology. GB 34 (*yáng líng quán*), and SP 9 (*yīn líng quán*) can be added as supplementary points; add warming needle moxibustion or mild moxibustion when the pattern presents with signs of wind *bì*, painful *bì* or fixed *bì*. For heat *bì* , needle with drainage; moxibustion is not applicable. For patients with inhibited movement, use electro-acupuncture with an intermittent wave setting. Blood injection therapy at ST 36 (*zú sān lĭ*), GB 34 (*yáng líng quán*) and SP 10 (*xuè hăi*) also can be employed as a daily treatment.

4. Bee Venom Therapy

Bee venom therapy, also known as apitherapy, is an integration of a bee venom sting therapy and traditional Chinese medicine where the stinger of the bee is inserted into selected acupoints. The pharmacological action of the bee venom in combination with the effect of acupoint stimulation makes this a commonly used clinical therapy in China.

【Prescription】

The same as for standard needling.

【Indications】

All types of knee pain.

【Manipulations】

① Preparation: Remove bees from the hive and place them in a small box.

② Skin test: Before administration of the bee venom therapy, a bee venom allergy test should be applied. For those seriously allergic to the venom, other methods should be considered. If necessary, the venom should first be deallerginizied before treatment.

On the patient's first visit, a bee venom allergy test is conducted on the skin of the back. After local skin disinfection, take one bee and sting the skin on one side of the spinal column of the back, and withdraw the stinger immediately. If there is no obvious reaction, on the next day, sting once again on the skin of the other side and withdraw the stinger after one half of a minute. Observe both local and general reactions, blood pressure, and temperature. If there is slight local reaction in half an hour, no obvious swelling, or the diameter of the swelling is less than 5 cm, or no obvious changes of the temperature and the blood pressure, this indicates a negative negative response, and venom therapy can be applied.

③ Select the acupoints and apply routine disinfection.

④ Lightly grasp the thorax or the lumbar region of the bee with a pair of tweezers, and with the tail contacting the acupoint, insert the stinger into the skin. Remove the bee and retain the stinger. With the toxin at the bee's stinger contracting intermittently, the bee venom is transmitted throughout the whole body. The stinger is generally retained in the skin for 15 minutes.

⑤ Withdraw the stinger with a pair of tweezers and observe the patient for 30 minutes in case of allergic reaction.

Initially, 1 to 2 bees can be used each time. For those without any serious reaction, add one bee sting each time and according to the situation and the patient's reaction decide the total number of the bees to be used. No more than 10 bees should be used during a treatment. Apply once every day with 3 to 5 treatments making up one course of treatment.

【Notes】

① Before performing bee venom therapy, the skin allergy test should be applied.

② After bee venom therapy some patients may have certain reactions such as fever, local itching and swelling, lymph node enlargement, etc. For those with slight reactions, just drinking more water will help these reactions disappear over time. For those with more obvious reactions, decrease the number of the bees used or prolong the interval period between treatments.

③ During bee venom therapy, the condition of the patient should be carefully monitored. If there are any serious allergic reactions or allergic shock, seek medical attention immediately and send the patient to the hospital.

【Emergency Treatment of Allergic Reactions】

① Withdraw the bee stinger immediately; do not squeeze the white toxicyst.

② For patients in shock, immediately administer muscular or subcutaneous injection of 1:1000 adrenaline, 0.3 to 0.5 ml. If the blood pressure doesn't ascend or still remains at the shock level, repeat the injection every 15 to 30 minutes until the blood pressure ascends over the shock level and remains stable.

③ Muscular injection of diphenhydramine 25 mg.

④ For those with trouble breathing, asthma, laryngeal edema or hoarseness, use steam inhalation of isoprenaline, salbutamol, or sublingual administration of isoprenaline 10 mg, or oral administration of ephedrine 25mg.

⑤ For those with difficulty breathing or apnea, use oxygen inhalation; for those with serious laryngeal edema without any immediate improvement after drug administration, perform a tracheotomy.

⑥ Intravenous Drip: Using 10% glucose liquid, Vitamin C 1-2 g plus hydrocortisone 200-300 mg or dexamethasone 5-10 mg. According to the clinical situation, norepinephrine, coraminum, lobeline or caffeine can be selected. Additionally, intravenous injection with 25-50% glucose liquid plus hydrocortisone 50-100 mg can be used, or muscular injection of dexamethasone 5 mg, intravenous injection of calcium gluconate and Vitamin C are also acceptable.

⑦ For those with obvious local swelling and pain, apply ice water locally with a towel, or apply an ointment such as dexamethasone compound or acetate cream, or use Jidesheng brand Sheyao tablets prepared with cold boiled water.

⑧ For those with rash, itching, or edema, oral administration of chlorpheramine maleate a 4 mg or hydroxyzine 25 mg.

Since the skin test is applied in advance, those patients seriously allergic to bee venom are excluded from treatment. Having engaged in the clinical bee venom therapy for many years, the author and other researchers have not encountered any serious allergic reactions, indicating that as long as the skin test is strictly applied, bee venom therapy is generally safe.

5. Abdominal Acupuncture

Abdominal acupuncture follows the theory of TCM as its foundation, with the *shén què* regulatory system at its core. The theory of abdominal acupuncture states that a human's prenatal condition, from formless essential qi to the formation of the embryo, totally depends on the *shén què* regulatory system. So the *shén què* regulatory system is perhaps the earliest system of the embryonic stage, and it may also be the mother of the channel systems which regulate the body.

Abdominal acupuncture is established on multiple levels of the channel's spatial structure and changes. After over 20 years of experimentation on the effects of needling depth, abdominal acupuncture has summarized the related rules of needling according to the depth of the illness, and also avoids the conventional concept of the "arrival of qi"; this method advances a new acupuncture concept of "needling at the location of disease". The principle for devising an acupuncture formula can be divided into choosing a principal point, a supplementary point, an assistant point, and an envoy point (chief, deputy, assistant and envoy).

KOA is always caused by obstruction of qi and blood due to *ying-wei* weakness, deficient sinew and bone, and the local attack of wind and cold.

【Prescription】

ST 24 (*huá ròu mén*) (on the affected side), ST 26 (*wài líng*) (on the affected side), *qì páng* (on the non-affected side), and *xià fēng shī diǎn* (on the affected side).

【Indications】

The same as for standard needling.

【Manipulations】

The patient assumes a supine position with the abdomen exposed. Before needling, the palpate the whole abdomen in order to check for swelling of the liver and spleen and for tenderness or masses. The therapy can be applied to the patient without positive signs. After routine skin disinfection, select a 40-50 mm needle according to the body of patient, and then insert the needle perpendicularly into the point. Instead of lifting and thrusting, light manipulations such as the twirling method are employed; there is no need to obtain the sensation of soreness, numbness, distention and pain, but it will be more effective when there is a sensation of heaviness and tightness under the finger, like a fish nibbling on bait.

6. Point Application Therapy

Point application therapy is an external TCM treatment that refers to placing a plaster made up of a variety of medicinal powders on certain points or around the affected area to prevent and treat disease by means of skin absorption and the function of the channels.

【Prescription】

The same as for standard needling

【Indications】

Patterns of wind-cold-damp *bì*, phlegm stasis and obstruction, liver and kidney deficiency, and cold congealing due to yang deficiency.

【Preparation】

Prepare *wēi líng xiān* (威灵仙, Radix et Rhizoma Clematidis, Chinese clematis root), and *cōng bái* (葱白, Bulbus Allii Fistulosi, scallion), and an appropriate amount of vinegar.

7. Auricular Acupuncture

Auricular acupuncture is one of the acupuncture therapies used to prevent and treat diseases. It is applied to the ear by needling, bloodletting, massage, plasters, and injection.

The ear is not an isolated organ in TCM; on the contrary, it connects with the *zang-fu* organs physiologically and pathologically. The *Yellow Emperor's Inner Classic* and *Spiritual Pivot* (*Huáng Dì Nèi Jīng Líng Shū*, 黄帝内经灵枢) records that: "the qi and blood of the twelve channels and 365 collaterals traverse upward to the facial area and the orifices. The divergent qi passes through the ear and transforms into hearing. The qi and blood of the twelve channels and 365 points flows upward and penetrates into the five sensory organs and seven orifices, and their divergent qi and blood penetrates into the ear. This confirms that the ear and the body are a united and indivisible whole, which is the theoretical basis for diagnosing and treating diseases with auricular acupuncture.

【Prescription】

Shen men (*shén mén*, 神门), Adrenal gland (*shèn shàng xiàn*, 肾上腺), Liver (*gān*, 肝), tender points on the knee.

【Indications】

Patterns of wind-cold-damp *bì*, phlegm stasis and obstruction, liver and kidney deficiency, cold congealing due to yang deficiency.

【Manipulations】

Method Ⅰ: Auricular Needling

After disinfection, use intense stimulation, or retain the needles on the two ears alternately.

Method Ⅱ: Auricular Plaster

Wáng bù liú xíng (王不留行, Semen Vaccariae) is fastened to the ear with a piece of adhesive tape; press and apply the plaster simultaneously. The patient is recommended to press and manipulate the selected points 3-5 times each day.

Auricular therapy is easily applied and is well tolerated by patients. It is effective and worthy to be spread due to its simple application and wide indications.

8. Fire Needling Therapy

Fire needling is one of therapies in TCM. Its point selection is the same used as in the filiform needle method, and is applicable for patterns of wind-cold-damp *bì* and cold congealing due to yang deficiency. Disinfect the area to be treated, heat the needle with an alcohol burner, quickly insert the needle, and then withdraw immediately. Applicable for the four limbs, lumbus, and abdominal regions.

9. Collateral Bloodletting Therapy

Collateral bloodletting therapy is a treatment in which certain acupoints or superficial vessels are pricked and bled. It is also called the "vessel-pricking therapy" or "collateral-pricking therapy". This therapy is usually applied on *ashi* points or superficial veins and its indication is for heat *bì*.

Note: Fire needling and collateral bloodletting therapy are not allowed in certain countries, so only a brief introduction is provided here.

Tuina

Tuina therapy refers to a variety of manipulations that can be applied to the knee joint and also passive movements that act to unblock the channels, warm the channels and dissipate cold, relax and remove adhesions, and promote the recovery of soft tissue. *Tuina* has a long history with many treatment applications for KOA.

1. Block-Resolving Manipulation

When something is blocked between the joint spaces, no matter where it is, the loosened body structure or meniscus will fracture; this can produce severe pain and dysfunction. Such blockages should be quickly resolved to relieve pain.

① The patient assumes a supine position and lifts the affected joint. The assistant supports and fixes the thigh of the affected leg. The physician holds the ankle with one hand and creates traction, rotation, and shakes and extends the leg simultaneously. Inwardly press the tendor spot on the joint interspace with the other hand until the knee can extend straight and move without inhibition. Most of the symptoms can be resolved by block-resolving manipulation. If the patient is obese, the physician can hold the ankle with his arm, and rotate and flex the lower leg with his hand.

② The patient is in the supine position and the physician lifts the affected joint. The physician shoulders the patient's knee and faces to the ankle side, pulling back on the thigh, producing traction, rotation and extension with both hands. This technique can resolve the block after a few treatments.

③ Pushing and Pulling Reduction: The patient lying in the supine position, the patient keeps his affected knee at a 90 degree angle. The physician stands on the affected side of the patient, seats himself or places his knee on the patient's foot of the affected leg in order to fix the leg in position. Then embraces the upper part of the leg, pushes it forward and pulls it backward for several times (similar to the drawer test) or induces it to rotate medially and laterally at the same time.

2. Thumb-Pushing and Kneading Manipulation

The patient takes a supine or sitting position while the physician stands on the lateral side of the affected leg. Fix the leg with one hand, push and knead with the other hand on the pre-capsule of the knee joint, bilateral collateral ligament,

ligament of the patella, and post-capsule of the popliteal fossa. The finger force commences from light to heavy until local soreness and distention is achieved. Treatment is conducted once a day, 5-10 minutes per session, with 10 sessions completing one therapeutic course.

3. Tendon-Flicking Manipulation

The patient is in a supine or sitting position. The physician lays the thumb of the right hand on the lateral side of the knee joint, and the other four fingers on the medial corresponding side. First, flick the lateral tendon from outside to inside several times, next flick the medial tendon from inside to the outside several times. Finally, put the right hand beside the knee joint, and flick the tendon on the popliteal fossa several times. Treatment is rendered once a day, 30-60 minutes each time.

4. Patella-Pinching and Pushing Manipulation

The patient is in a sitting position. The physician uses the thumbs with the index fingers of the two hands to pinch and push the patella with opposite force. First, push the patella transversely, and then longitudinally and finally pushes it in a circular motion. This manipulation is given once a day, 20-30 minutes each time. Ten days forms one course of treatment.

5. Digital Pressing Manipulation

Grip and squeeze the bilateral interspace of the knee joint with the thumb, forefinger and middle finger for 1-2 minutes. Then press *nèi xī yǎn* (内膝眼), ST 35 (*dú bí*), the lower extremity of patella, *hè dǐng* (鹤顶), SP 10 (*xuè hǎi*), ST 34 (*liáng qiū*) and GB 31 (*fēng shì*); add an extra 2 minutes of treatment at the most tender spot. Treat two times a month for 20-30 minutes each time. Twenty sessions form one therapeutic course of treatment.

6. Flexing and Stretching Manipulation

With the patient in a supine position, the physician supports the affected knee with one hand, holds the ankle with the other hand to force the knee joint to be stretched to its maximal limitation, and retains it there for several seconds or vibrates the leg slightly several times; relax and repeat the manipulation 1-2 more times.

With the patient in a supine position, the physician holds the posterior side of the thigh with one hand, supports the ankle with the other hand to flex the knee joint to the barrier position and retains it there for several seconds; then relax and repeat another 1-2 times. Apply 2-3 times in one week, 10-15 minutes for each session. 10 sessions form one therapeutic course of treatment with an interval of 7 days between courses.

7. Tendon Releasing Manipulation

With the patient in a supine position, the physician first applies grasping and rolling techniques to relax the affected muscles. An assistant holds the lower part of the thigh, while the physician supports the ankle to cause opposite traction. After that, the physician initiates flexing, stretching and rotating manipulations, and then repeats the procedure 1-2 times. At last, grasping, kneading and patting of the affected leg completes the therapy. Treat every other day, with ten days forming one course of treatment.

Chinese Medicinal Formulas

1. Fáng Fēng Tāng (*Ledebouriella Decoction*, 防风汤)

【Corresponding Patterns】

Migratory *bì*

【Prescription】

防风	*fáng fēng*	15 g	Radix Saposhnikoviae
秦艽	*qín jiāo*	15 g	Radix Gentianae Macrophyllae
麻黄	*má huáng*	6 g	Herba Ephedrae
杏仁	*xìng rén*	15 g	Semen Armeniacae Amarum
葛根	*gé gēn*	15 g	Radix Puerariae Lobatae

赤茯苓	*chì fú líng*	15 g	Poria Rubra
当归	*dāng guī*	10 g	Radix Angelicae Sinensis
肉桂	*ròu guì*	10 g	Cortex Cinnamomi
生姜	*shēng jiāng*	10 g	Rhizoma Zingiberis Recens
大枣	*dà zăo*	10 g	Fructus Jujubae
甘草	*gān căo*	6 g	Radix et Rhizoma Glycyrrhizae
黄芩	*huáng qín*	10 g	Radix Scutellariae

【Formula Analysis】

Fáng fēng (Radix Saposhnikoviae, siler), and *qín jiāo* (Radix Gentianae Macrophyllae, large leaf gentian root) can dispel wind and eliminate dampness, unblock the collaterals and relieve pain.

Má huáng (Herba Ephedrae, ephedra herb), and *xìng rén* (Semen Armeniacae Amarum, almond) have the function of dissipating cold, diffusing the lung, and venting the pathogen outward.

Gé gēn (Radix Puerariae Lobatae, kudzuvine root) releases the flesh and expels pathogens *chì fú líng* (Poria Rubra, Indian bread pink epidermis) percolates and drains dampness with bland medicinals.

Dāng guī (Radix Angelicae Sinensis, Chinese angelica) can nourish and invigorate the blood; *ròu guì* (Cortex Cinnamomi, cassia bark) is able to warm yang.

Shēng jiāng (Rhizoma Zingiberis Recens, fresh ginger) and *dà zăo* (Fructus Jujubae, Chinese date) are used to harmonize and regulate the center and *ying*.

Gān căo (Radix et Rhizoma Glycyrrhizae, liquorice root) unblocks the collaterals, soothes the joints and harmonizes the medicinals.

Huáng qín (Radix Scutellariae, scutellaria root) is applied to prevent excessive acrid-warming medicinals from transforming into fire that damages blood and consumes qi.

【Modifications】

➢ Add *fù pià* (附片, Radix Aconiti Lateralis Praeparata, processed aconite root slices) for the condition of severe pain aggravated by cold.

➢ Add *niú xī* (牛膝, Radix Achyranthis Bidentatae, two-toothed achyranthes root) and *sāng jì shēng* (桑寄生, Herba Taxilli, Chinese taxillus herb) for knee joint degeneration.

2. *Wū Tóu Tāng (Radix Aconiti Praeparata Decoction, 乌头汤)*

【Corresponding Patterns】

Painful *bì*

【Prescription】

制川乌	*zhì chuān wū*	6 g	Radix Aconiti Praeparata
麻黄	*má huáng*	10 g	Herba Ephedrae
芍药	*sháo yào*	15 g	Radix Paeoniae
甘草	*gān cǎo*	6 g	Radix et Rhizoma Glycyrrhizae
蜂蜜	*fēng mì*	10 g	Mel
黄芪	*huáng qí*	25 g	Radix Astragali

【Formula Analysis】

Zhì chuān wū (Radix Aconiti Praeparata, prepared monkshood mother root), and *má huáng* (Herba Ephedrae, ephedra herb) can warm the channels and dissipate cold, unblock the collaterals and relieve pain.

Sháo yào (Radix Paeoniae, peony root) and *gān cǎo* (Radix et Rhizoma Glycyrrhizae, liquorice root) relax the spasms.

Huáng qí (Radix Astragali, milk-vetch root) is used to boost qi, consolidate the exterior, promote blood flow, and unblock qi.

【Modifications】

➢ Add *fù zǐ* (附子, Radix Aconiti Lateralis Praeparata, monkshood), *xì xīn*

(细辛, Radix et Rhizoma Asari), *guì zhī* (桂枝, Ramulus Cinnamomi), *gān jiāng* (干姜, Rhizoma Zingiberis), and *dāng guī* (当归, Radix Angelicae Sinensis,) for cold joints and severe pain aggravated by cold.

➢ Replace *zhì chuān wū* (制川乌, Radix Aconiti Praeparata) with *chuān wū* (川乌, Radix Aconiti) or *cǎo wū* (草乌, Radix Aconiti Kusnezoffii) for excess cold and dampness.

3. *Yì yǐ rén Tāng* (*Semen Coicis Decoction,* 薏苡仁汤)

【Corresponding Patterns】

Fixed *bì*

【Prescription】

薏苡仁	*yì yǐ rén*	30 g	Semen Coicis
桂枝	*guì zhī*	15 g	Ramulus Cinnamomi
制川乌	*zhì chuān wū*	6 g	Radix Aconiti Praeparata
苍术	*cāng zhú*	15 g	Rhizoma Atractylodis
甘草	*gān cǎo*	6 g	Radix et Rhizoma Glycyrrhizae
羌活	*qiāng huó*	10 g	Rhizoma et Radix Notopterygii
独活	*dú huó*	10 g	Radix Angelicae Pubescentis
防风	*fáng fēng*	10 g	Radix Saposhnikoviae
麻黄	*má huáng*	10 g	Herba Ephedrae
川芎	*chuān xiōng*	10 g	Rhizoma Chuanxiong
当归	*dāng guī*	15 g	Radix Angelicae Sinensis

【Formula Analysis】

Yì yǐ rén, *guì zhī* and *zhì chuān wū* warm the channels and dissipate cold.

Cāng zhú and *gān cǎo* fortify the spleen and eliminate dampness.

Qiāng huó, *dú huó* and *fáng fēng* dispel wind and eliminate dampness.

Má huáng and *guì zhī* eliminate dampness and relieve pain.

Dāng guī and *chuān xiōng* nourish the blood and unblock the channels.

【Modifications】

➢ Add *bì xiè* (萆薢, Rhizoma Dioscoreae Hypoglaucae) and *mù tōng* (木通, Caulis Akebiae) to promote urination and unblock the channels in cases of dysuria or numbness of the skin.

➢ Add *hǎi tóng pí* (海桐皮, Cortex Erythrinae) and *xī xiān cǎo* (豨莶草, Herba Siegesbeckiae) to remove edema.

➢ Add *fú líng* (茯苓, Poria), *zé xiè* (泽泻, Rhizoma Alismatis), *chē qián zǐ* (车前子, Semen Plantaginis, plantago seed) to promote urination and dispel dampness.

➢ Add *bàn xià* (半夏, Rhizoma Pinelliae) and *tiān nán xīng* (天南星, Rhizoma Arisaematis) for phlegm-damp exuberance.

➢ Add *Juān Bì Tāng* (Eliminating *Bì* Decoction, 蠲痹汤) to boost qi, harmonize ying, dispel wind, overcome dampness, unblock the collaterals, and relieve pain. This formula can be used as the base formula for chronic *bì* patterns without an obvious predominance of wind-cold-dampness, or it can be modified according to the predominant external pathogen and symptoms.

4. *Bái Hǔ Jiā Guì Zhī Tāng* (*White Tiger Decoction Plus Cinnamon Twig*, 白虎加桂枝汤)

【Corresponding Patterns】

Heat *bì*

【Prescription】

石膏	*shí gāo*	20 g	Gypsum Fibrosum
知母	*zhī mǔ*	15 g	Rhizoma Anemarrhenae
桂枝	*guì zhī*	15 g	Ramulus Cinnamomi

甘草	*gān cǎo*	6 g	Radix et Rhizoma Glycyrrhizae
粳米	*jīng mǐ*	15 g	Oryza Sativa L

【Formula Analysis】

Bái Hǔ Tāng (White Tiger Decoction, 白虎汤) clear heats and nourishes yin. *Guì zhī* acts to scatter wind, release the exterior, and unblock the channels.

【Modifications】

➢ Add *mǔ dān pí* (牡丹皮, Cortex Moutan), *chì sháo* (赤芍, Radix Paeoniae Rubra, red peony root), *shēng dì* (生地, Radix Rehmanniae), and *zǐ cǎo* (紫草, Radix Arnebiae) in cases of erythema to clear heat, cool and invigorate blood, and dissolve stasis.

➢ Add *jīng jiè* (荆芥, Herba Schizonepetae), *bò he* (薄荷, Herba Menthae), *niú bàng zǐ* (牛蒡子, Fructus Arctii) and *jié gěng* (桔梗, Radix Platycodonis) in cases with fever, aversion to wind, or sore throat.

➢ Add *xuán shēn* (玄参, Radix Scrophulariae), *mài dōng* (麦冬, Radix Ophiopogonis), and *shēng dì* (生地, Radix Rehmanniae) for patterns of excess heat damaging yin with symptoms of thirst and vexation.

➢ *Wǔ Wèi Xiāo Dú Yǐn* (Five Ingredients Toxin-Removing Beverage, 五味消毒饮) combined with *Xī Huáng Wán* (Rhinoceros Bezoar Pill, 犀黄丸) can clear heat and resolve toxins, cool the blood and relieve pain. It can also be used in cases of inflamed joints with severe cutting like pain and heat sensation on palpation, spasms which are aggravated in the evening, high fever, vexation and thirst. Fire transforming from intense heat toxin penetrating deep into the joints and damaging the fluids cause the the tongue to be red with little fluids, and the pulse to be wiry and rapid.

5. *Shuāng Hé Tāng* (*Two Combination Decoction*, 双合汤)

【Corresponding Patterns】

Turbid phlegm and blood stasis.

【Prescription】

桃仁	*táo rén*	20 g	Semen Persicae
红花	*hóng huā*	20 g	Flos Carthami
当归	*dāng guī*	15 g	Radix Angelicae Sinensis
川芎	*chuān xiōng*	15 g	Rhizoma Chuanxiong
白芍	*bái sháo*	15 g	Radix Paeoniae Alba
茯苓	*fú líng*	15 g	Poria
半夏	*bàn xià*	15 g	Rhizoma Pinelliae
陈皮	*chén pí*	15 g	Pericarpium Citri Reticulatae
白芥子	*bái jiè zǐ*	10 g	Semen Sinapis
竹沥	*zhú lì*	10 g	Succus Bambusae
姜汁	*jiāng zhī*	10 g	Succus Rhizomatis Zingiberis

【Formula Analysis】

Táo rén, *hóng huā*, *dāng guī*, *chuān xiōng* and *bái sháo* invigorate blood, dissolve stasis, unblock channels, and relieve pain.

Fú líng, *bàn xià*, *chén pí*, *bái jiè zǐ*, *zhú lì* and *jiāng zhī* fortify the spleen and dissolve phlegm.

【Modifications】

➢ Add *dǎn nán xīng* (胆南星, Arisaema cum Bile, bile arisaema) and *tiān zhú huáng* (天竺黄, Concretio Silicea Bambusae, bamboo sugar) for retained phlegm and turbid subcutaneous nodules.

➢ Add *é zhú* (莪术, Rhizoma Curcumae, curcumae rhizome), *sān qī* (三七, Radix et Rhizoma Notoginseng, pseudoginseng root), *tǔ biē chóng* (土鳖虫, Eupolyphaga seu Steleophaga, ground beetle) for obvious blood stasis, painful swelling, and stiff or deformed joints with inhibited movement.

- Add *chuān shān jiǎ* (穿山甲, Squama Manitis, pangolin scales), *bái huā shé* (白花蛇, Agkistrodon, agkistrodon), *quán xiē* (全蝎, Scorpio, scorpion), *wú gōng* (蜈蚣, Scolopendra, centipede), and *dì lóng* (地龙, Pheretima) for binding of phlegm and stasis with excessive pain.
- Add *huáng bǎi* (黄柏, Cortex Phellodendri Chinensis) and *mǔ dān pí* (牡丹皮, Cortex Moutan) for phlegm stasis transforming into internal heat.

6. *Bǔ Xǔe Róng Jīn Tāng* (*Blood-Supplementing Sinew-Nourishing Decoction*, 补血荣筋汤)

【Corresponding Patterns】

Liver and kidney deficiency

【Prescription】

熟地黄	*shú dì huáng*	15 g	Radix Rehmanniae Praeparata
肉苁蓉	*ròu cōng róng*	15 g	Herba Cistanches
五味子	*wǔ wèi zǐ*	15 g	Fructus Schisandrae Chinensis
鹿茸	*lù róng*	15 g	Cornu Cervi Pantotrichum
菟丝子	*tù sī zǐ*	15 g	Semen Cuscutae
牛膝	*niú xī*	10 g	Radix Achyranthis Bidentatae
杜仲	*dù zhòng*	10 g	Cortex Eucommiae
桑寄生	*sāng jì shēng*	10 g	Herba Taxilli
天麻	*tiān má*	10 g	Rhizoma Gastrodiae
木瓜	*mù guā*	10 g	Fructus Chaenomelis

【Formula Analysis】

Shú dì huáng, *ròu cōng róng*, and *wǔ wèi zǐ* enrich yin, supplement the kidney, nourish blood, and warm the liver.

Lù róng, tù sī zǐ, *niú xī*, and *dù zhòng* supplement liver and kidney and strengthen bone and muscle.

Sāng jì shēng, *tiān má*, and *mù guā* dispel wind and dampness, relax the tendons, unblock the channels, and relieve pain.

【Modifications】

➢ Add *lù jiăo shuāng* (鹿角霜, Cornu Cervi Degelatinatum), *xù duàn* (续断, Radix Dipsaci), and *gŏu jĭ* (狗脊, Rhizoma Cibotii) for cases with kidney deficiency, lassitude of the loin and knee and hypodynamia.

➢ Add *fù zĭ* (附子, Radix Aconiti Lateralis Praeparata), *gān jiāng* (干姜, Rhizoma Zingiberis), and *bā jĭ tiān* (巴戟天, Radix Morindae Officinalis), or combine with *Yáng Hé Tāng* (Harmonious Yang Decoction, 阳和汤) for yang deficiency with cold limbs, aversion to cold, and joint spasms.

➢ Add *guī băn* (龟板, Plastrum Testudinis), *shú dì huáng* (熟地黄, Radix Rehmanniae Praeparata), *nǚ zhēn zĭ* (女贞子, Fructus Ligustri Lucidi), or combine with *Hé Chē Dà Zào Wán* (Placenta Hominis Recovery Pill, 河车大造丸) for liver and kidney yin deficiency with painful knees and loins, hypothermia and vexation, or afternoon tidal fever.

➢ *Dú Huó Jì Shēng Tāng* (Pubescent Angelica and Mistletoe Decoction, 独活寄生汤) can be used in all chronic *bì* syndromes manifesting with a pale complexion, weak breathing, spontaneous sweating and fatigue, muscular atrophy, lassitude of the loins and legs, vertigo and tinnitus caused by pathogen retention due to deficient healthy qi, and insufficient qi and blood. It acts to supplement liver and kidney, nourish qi and blood, dispel wind, remove dampness, and eliminate obstruction to activate the collaterals.

Other Therapies

1. Medicinal Compress Therapy

Besides oral medications, medicinal compress therapy can be used alone or be combined with acupuncture and *tuina* manipulation.

【Indications】

The same as point application therapy.

【Prescription】

To dispel wind and unblock the collaterals add *wū shāo shé* (乌梢蛇, Fructus Mume), *dì lóng* (地龙, Pheretima), *wēi líng xiān* (威灵仙, Radix et Rhizoma Clematidis), *chuān wū* (川乌, Radix Aconiti), *cǎo wū* (草乌, Radix Aconiti Kusnezoffii), *shēn jīn cǎo* (伸筋草, Herba Lycopodii), and *tòu gǔ cǎo* (透骨草, Herba Vaccinii Urophylli).

For patterns of cold and dampness add *fáng fēng* (防风, Radix Saposhnikoviae), *qiāng huó* (羌活, Rhizoma et Radix Notopterygii), *niú xī* (牛膝, Radix Achyranthis Bidentatae), *fù zǐ* (附子, Radix Aconiti Lateralis Praeparata), *gān jiāng* (干姜, Rhizoma Zingiberis), *ròu guì* (肉桂, Cortex Cinnamomi), and *bīng piàn* (冰片, Folium Artemisiae Argyi).

For heat patterns add *jīn yín huā* (金银花, Flos Lonicerae Japonicae), *tǔ fú líng* (土茯苓, Rhizoma Smilacis Glabrae), *fáng jǐ* (防己, Radix Stephaniae Tetrandrae).

For liver and kidney insufficiency patterns add *bǔ gǔ zhī* (补骨脂, Fructus Psoraleae), *dù zhòng* (杜仲, Cortex Eucommiae), *sāng jì shēng* (桑寄生, Herba Taxilli), *lù jiǎo shuāng* (鹿角霜, Cornu Cervi Degelatinatum), *xiān líng pí* (仙灵脾, Herba Epimedii), and *liú jì nú* (刘寄奴, Herba Artemisiae Anomalae).

In addition, to invigorate the blood and relieve pain add *tián qī* (田七, Radix et Rhizoma Notoginseng), *dāng guī* (当归, Radix Angelicae Sinensis), *yán hú suǒ* (延胡索, Rhizoma Corydalis), *chì sháo* (赤芍, Radix Paeoniae Rubra), *hóng huā* (红花, Flos Carthami, Safflower), *táo rén* (桃仁, Semen Persicae), *jī xuè téng* (鸡血藤, Caulis Spatholobi,), *xuè jié* (血竭, Sanguis Draconis), *rǔ xiāng* (乳香, Olibanum, frankincense), and *mò yào* (没药, Myrrha, myrrh).

All the medicinals combined act to improve blood circulation and promote absorption of inflammatory substances by invigorating blood and dissolving stasis, relaxing the sinews, quickening the collaterals, dissipating swelling and cold, soothing joint motion, and releasing adhesions.

【Manipulations】

Grind and sift all medicinals into powder and preserve them appropriately.

When the therapy is going to be applied, take out a proper amount of powder and place into the water, and boil with vinegar. Then mix them into a pasteand place into a cloth gauze bag; apply the bag to the affected knee joint while hot.

The powder can also be mixed with honey into a paste which is placed onto a cotton cushion and applied directly on the affected knee joint; or, put the powder into a cloth bag and steam it in a pan; apply when the temperature can be tolerated. Therapy is given 2-3 times a day, 20-30 minutes each time, and continued for 10-20 days. Stop for 2-3 days between courses.

Medicinal compress application is a therapy with a two-sided curative effect which takes advantage of heat energy and the medicinal potency. The medicinals act to invigorate blood, unblock the collaterals, dispel wind, and eliminating dampness, while the heat energy makes the skin surface congest and expand, which improves absorbtion of the medicinals into the body while improving local blood circulation. In clinical practice, the medicinal effect lasts for 2-3 days; this is inferior to standard medications, but still there is confirmed efficacy for dispersing swelling and improving mobility.

Medicinal compress application therapy has clear advantages, such as the potency of the medicinals going directly to the affected area with few side effects.

2. Medicinal Steaming and Fumigation Therapy

Medicinal steaming and fumigation therapy is an effective external therapy that refers to fumigation over certain areas, acupoints, channels and organs with medicinal steam for the purpose of treatment, rehabilitation and fitness.

Treatise on Cold Damage (*Shāng Hán Lùn*, 伤寒论) states: "yang qi constrained at the exterior is supposed to be steamed and fumigated".

Secret Formulary Bestowed by Immortals for Treating Injuries and Mending Fractures (*Xiān Shòu Lǐ Shāng Xù Duàn Mì Fāng*, 仙授理伤续断秘方) has also recorded that in ancient times, hot compresses and fumigation therapy were called "*lín tuò*", "*lín xǐ*" or "*lín yù*". By boiling the medicinals in a pan and fumigating the affected area with the steam, the medicinal potency is better able to penetrate deeply into the affected areas, acting to unblock the joints, dredge the interstitial spaces, invigorate blood, dissolve stasis, rectify qi, relieve pain, and release spasms and swelling. This is accomplished by means of the coordinated functional regulation of the combined heat and the medicinal properties. In is the opinion of Western medicine, heat stimulation can improve blood circulation, aid in dispersing and absorbing hematomas, and also relieve spasms and stiffness.

【Indications】

The same as for the medicinal compress.

【Prescription】

TCM theory states that KOA is caused by channel damage, qi stagnation, blood stasis, and collateral obstruction, so its basic formula is composed of medicinals with the function of unblocking the channels, relaxing the sinews, dispelling wind and eliminating dampness. Medicinals include *shēn jīn cǎo* (伸筋草, Herba Lycopodii), *tòu gǔ cǎo* (透骨草, Herba Vaccinii Urophylli), *hǎi tóng pí* (海桐皮, Cortex Erythrinae), and *mù guā* (木瓜, Fructus Chaenomelis).

Other medicinals can be added to the formula to help dispel stasis and swellings in patterns of severe cold such as *sū mù* (苏木, Lignum Sappan), *niú xī* (牛膝, Radix Achyranthis Bidentatae), and *chuān xiōng* (川芎, Rhizoma Chuanxiong); add *chuān wū* (川乌, Radix Aconiti), *cǎo wū* (草乌,Radix Aconiti Kusnezoffii), *dú huó* (独活, Radix Angelicae Pubescentis), *qín jiāo* (秦艽, Radix Gentianae Macrophyllae) and *guì zhī* (桂枝, Ramulus Cinnamomi).

To further disspate stasis and reduce swelling add *rǔ xiāng* (乳香, Olibanum), *mò yào* (没药, Myrrha, myrrh), *jī xuè téng* (鸡血藤, Caulis Spatholobi), *sān léng* (三棱, Rhizoma Sparganii), and *é zhú* (莪术, Rhizoma Curcumae).

【Preparation】

Place one batch of the formula in an earthware pot with water in a low position and heat the contents gradually until boiling and steaming, then turn the heat down to a lower temperature. The patient then exposes the affected knee joint above the steam. Adjust the distance between the knee joint and the steaming pot until the proper temperature is achieved without scorching the patient.

Medicinal steaming and fumigation therapy achieves its potency by means of the temperature and the function of the medicinal formula. Steaming the knee joint over the pot is an effective physical treatment; as capillary vessels the affected area become dilated, blood circulation and metabolism is increased and tissue nourishment and function are improved. Edema and inflammatory substances are absorbed by steaming and fumigation and osmosis of the medicinals stimulate the peripheral nerves thereby relieving pain, decreasing tension and stiffness of the local tissue and relaxing the muscles and ligaments. This effect results in improved movement of the limbs and joints, allowing them to move freely and recovery more rapidly.

3. Intra-Articular Injection Therapy

Regardless of whether treatment is rendered in an outpatient or inpatient facility, intra-articular injection therapy is acommonly used procedure. Intra-articular injection can help control the degeneration of the articular cartilage. Compared to internal or external medicinal treatments, intra-articular injection is a completely different method of treatment.

Treatment is applied directly to the affected knee cavity.

【Indication】

Knee osteoarthriris (KOA)

【Prescription】

The medicines used for intra-articular injection therapy include *danshen* root injection, *chuanxiong* root injection, piperazine injection, chitosan, hormones, and hyaluronic acid.

【Procedure】

The patient assumes a sitting or supine position with the knee flexed. Sterilize the area to be treated and specifically puncture one of the four access points: the medial or lateral side of the superior patella, or the medial or lateral side of the inferior patella.

4. Acupotomy

Acupotomy, an occlusive operation, integrates traditional acupuncture with modern surgery.

【Indications】

Intractable pain, hyperosteogeny, bursitis, morbidity of the muscles and ligaments

【Manipulation】

According to the pathological changes, insert the needle-shaped knife at the border of the hyperosteogeny or bone spur, make an incision and attempt to remove the bone spur using shoveling and grinding manipulations to smooth down or remove the spur.

5. Point Injection Therapy

Point injection therapy, also called hydro-acupuncture, refers to the injection of medicinals or drugs into acupoints.

【Indications】

All *bì* patterns affecting the knee

【Prescription】

Compound angelica injection, dan shen root injection, Chinese clematis root injection, wild Chinese quince fruit injection, paniculate swallowwort root injection, milk-vetch root injection, or vitamin B_1, vitamin B_{12}, Calcium Colloidal et Vit D, Plancenta Histolysate.

【Procedure】

Sterilize the local area and rapidly inject the needle into the point with a sterilized syringe until qi arrives. Then, pull the syringe back and if there is no blood in the syringe, inject the medicinal or drug.

Note: Intra-articular injection therapy, acupotomy and point injection therapy are three types of special therapies that can produce instant results, but in certain countries or areas they can not legally be used because of limitations placed on the scope of practice.

Case Studies

Mrs. Huang, age 62.

First visit: April, 2007

【Chief Complaint】

Repeated pain of both knee joints for ten years, aggravation accompanied with inhibited movement for one year.

【Medical History】

Ten years ago the patient suffered from pain in both knee joints without any evident injury. The pain was aggravated by overload and movement and was relieved with rest. The patient went to the hospital several times and was diagnosed with KOA. She was treated with oral medicines, physical therapy and intra-articular injection therapy. After the treatment the symptoms were relieved, but the pain relapsed repeatedly. One year ago, the pain was evidently aggravated by movement and was not relieved by oral medicine or physical therapy, moreover, the two knee joints were gradually inhibited when flexing and extending. In order to get a better effect, she asked for further treatment.

Clinical manifestations: stabbing pain of the knee joints, right side worse than the left. The pain was fixed, induced by cold, and aggravated by labor. The tissue around the knee joint was dark, swollen and stiff with inhibited flexion and extension when palpating. The patient had edema with a pale complexion, a preference for warm and an aversion to cold, chest distress and profuse phlegm. The tongue was a purple dusky color with a greasy white coating; the pulse was deep and choppy.

【Examination】

The two knee joints are swollen and deformed with obvious tenderness on the upper and lower extremities of patella. A sensation of friction and pain is produced when the patella is moved negatively. The floating patella test is negative. Tenderness is slight in the interspaces on the medial and lateral sides of the knee joint. Both lateral movement test and the drawer test are also negative. When the knee joint is stretched and flexed, there is obvious pain and friction of the knee can be felt as well.

【Diagnosis】

TCM diagnosis: knee *bì*

Western diagnosis: Knee Osteoarthritis (KOA)

【Pattern Differentiation】

Liver and kidney insufficiency, cold congealing due to yang deficiency, phlegm stasis and obstruction.

【Treatment Principles】

Supplement and boost the liver and kidney, unblock the channels and quicken the collaterals, warm yang, dissipate cold, dispel phlegm.

【Treatment】

Point Selection

ST 34 (*liáng qiū*)	SP 10 (*xuè hǎi*)	EX-LE5 (*xī yǎn*)
ashi points	ST 36 (*zú sān lǐ*)	BL 23 (*shèn shù*)
BL 17(*gé shù*)	BL 18 (*gān shù*)	RN 4 (*guān yuán*)
RN 6 (*qì hǎi*)		

Manipulations

Acupuncture: the patient assumes a supine position with both knee joints flexed.

ST 34 (*liáng qiū*), SP 10 (*xuè hǎi*), ST 36 (*zú sān lǐ*), and GB 34 (*yáng líng quán*) can be inserted 1-2 *cùn* perpendicularly with lifting, thrusting and twirling manipulations with drainage to produce local soreness and distention or a needle sensation spreading around the knee area or knee joint cavity.

Insert EX-LE5 (*xī yǎn*) 1-2 *cùn* obliquely, apply the even method from the anterior lateral side towards the posterior medial side. Soreness and distention should spread throughout the whole joint, or sometimes conduct downward.

Ashi points can be inserted with an approach of short needling, that is inserting the needle from a superficial level to a deep level and shaking the needle handle simultaneously until the needle arrives at bone, then pounding near the periosteum up and down like rubbing and scraping the bone.

Then apply warming-needle moxibustion and electro-acupuncture with an intermittent pulse for 15 minutes.

Therapy is given every other day, ten days constitutes one course of treatment.

Point Injection Therapy

Points A: bilateral BL 23 (*shèn shù*) and BL 18 (*gān shù*)

Points B: bilateral SP 10 (*xuè hǎi*) and BL 17(*gé shù*)

Medicinals: Angelica Compound Injection

Administration: alternately inject the points of group A and B with angelica compound injection. Use 4 points and 4 ml each time, using 1 ml for each acupuncture point. This therapy is given every other day, applied after filiform needle therapy. Ten treatments form one course of treatment.

Moxibustion

Instruct the patient to perform moxibustion on RN 4 (*guān yuán*), RN 6 (*qì hǎi*), and ST 36 (*zú sān lǐ*) for 10 minutes on each point, 1 to 2 times daily.

【Follow-up】

The pain was relieved after the first treatment. After insisting on this treatment for a whole therapeutic course, the knee joint movement was markedly improved and the swelling was dispersed, but there was still pain when initiating movement. In the second therapeutic course, collateral bloodletting and cupping therapies were applied once a week, combined with the other therapies mentioned above; therefore, a better effect was achieved. Two courses of treatment later, the two knee joints were able to move and the patient could walk freely without pain or inhibition.

【Discussion】

Root-deficiency with branch-excess as the foundation of KOA:

The internal etiology refers to liver and kidney deficiency, deficiency and failure of yang-qi to supply nourishment to the sinew and bone causing obstruction of blood and qi. Where there is malnutrition, there is pain. The *Chapter of Discussion on Pain* in *Basic Questions* states: "exhaustion of yin qi and dysfunction of yang qi result in sudden onset of pain".

Deficient healthy qi fails to promote, warm and nourish the *zang-fu* organs, channels and collaterals, sinew and bone, resulting in spasm and pain. "Blood governs moistening" refers to blood circulating within the vessels by which the blood is carried to the *zang-fu* organs internally and to the skin, muscles, tendons and bones externally. It circulates ceaselessly like a ring without an end to nourish and moisten all of the organs and tissues.

The *Chapter of Discussion on the Foundation-Zang-fu Organs* in *Ling Shu* states: "harmony of the blood is responsible for the ceasless flow of the channels, rehabilitation of ying-yin and yang, restores forceful sinew, bone and smooth joints". Blood deficiency fails to enrich, so the sinews will not be nourished and the vessels and collaterals become spasmodic. If the deficiency results in stasis, then pain will occur.

Branch-excess is the external etiology which refers to an attack of wind-cold-dampness along with the production of blood stasis and qi stagnation which can result in obstruction of the channels that inhibits the flow of qi and congeals blood. It is a fact that where there is blockage, there is pain.

The *Chapter of Discussion on Pain* in *Basic Questions* states: "attack of cold leads to stagnation and congealing flow of qi and blood, so if the cold attacks from outside the channel, it results in congealing and deficiency of blood; if the cold is retained in the channel, this results in blockage and produces pain".

So, cold congealing in the vessels leads to inhibited flow of qi and blood, and this produces pain, but dampness invasion can also block the flow of qi and blood to produce pain in the muscles and joints.

Discussion of the pathomechanism of KOA from the theory of collaterals:

The qi mechanism can be affected by the interaction of humans and nature involving external-contraction of the six pathogenic factors, internal damage caused by the seven emotions, traumatic injury, seasonal epidemics, and alcohol poisoning; can all cause disease. Often times after attack from any of these pathogenic factors, the flow of qi and blood becomes stagnant, so morbidity of the collaterals will occur over time.

The main reason is that the channels are thick while the collaterals are comparatively thin. The channel has an abundant amount of qi and blood and is not easily affected, while the collaterals are always thin and easily affected. In addition, qi deficiency increases the chance of infection; moreover, the pathogen lingers and easily produces morbidity of the collaterals.

It is confirmed that morbidity of the collaterals is mainly caused by a root-deficiency. First of all, the reason for collateral morbidity is root-deficiency and qi weakness, and next is the externally-contracted six pathogenic factors. For example, as wind-cold-dampness and heat pathogens are retained in the collaterals and battle with qi and blood for a long time, this produces stasis, congealing, collateral obstruction and morbidity.

In addition, other factors such as internal damage caused by the seven emotions, dietary impairment, and exhaustion or fatigue can damage the qi and blood of the collaterals. The *Chapter on Etiology of Disease* in the *Spiritual Pivot* states: "internal damage caused by grief and excessive thinking can make qi ascend and counterflow inhibiting the flow of six channels……congealing blood stagnates internally and is unable to be dispersed, body fluid is too dry to penetrate, so the pathogens are retained and can not be removed, therefore, abnormal masses form and come into being".

That is to say, excess of the five emotions can lead to counterflow of qi movement, inhibited blood flow, and engender stasis and obstruction of the collaterals. Excessive labor in ordinary life can damage yang energy and as insufficient yang qi fails to transport blood, so stasis *bì* forms. Immoderate eating and drinking, irregular living habits, and over straining can damage blood vessels and lead to bleeding. Exterior-interior obstruction of qi and blood due to internal damage by depression and anger can lead to the formation of abdominal masses.

Qi and yin damage and retained phlegm is commonly seen in the clinical setting. The obstructed collaterals inhibit the penetration of qi, blood and nutritive substances, resulting in malnutrition of the body and dysfunction of *zang-fu* organs. It is an absolute fact that a qi and blood deficiency condition will occur after a long period of illness. Obstructed collaterals also lead to extravasation of blood and external bleeding, which also consumes both qi and blood.

Prolonged obstructiuon of the collaterals is bound to transform into fire which consumes and scorches body fluids, thus damaging yin and blood. The collaterals are thinner than the channels, running in a zig-zag pattern all over the body. They have the function of penetrating and percolating body fluids and blood, balancing yin and yang, and interacting *ying-wei*. The nutritive supply of the collaterals itself also relies on the balanced composition of qi, blood, yin and yang. Prolonged channel and *zang-fu* disease can transmit to the collaterals and influence the transportation of qi, blood and body fluids, resulting in collateral morbidity.

Case Records as a Guide to Clinical Practice (*Lín Zhèng Zhǐ Nán Yī Àn*, 临证指南医案) states: "if a prolonged disease is over one hundred days, the vessels and collaterals must have become damaged".

The basic pathological changes of collateral morbidity are qi stagnation, phlegm coagulation and blood stasis. Collateral morbidity has a series of symptoms of deficiency-excess complex, but the general nature involves a deficiency of healthy qi.

Pain is a primary symptom of KOA and it includes various manifestations such as distending pain, stabbing pain, nocturnal pain, and intermittent pain. All of these result from collateral obstruction caused by qi stagnation, blood stasis, phlegm-dampness and external pathogens. Sinew binding, the characteristic symptom, is caused by obstruction of phlegm, stasis and dampness in the knee collaterals. The doctrine of collateral morbidity holds that "collateral obstruction is responsible for distention" and indeed distention and swelling is an important symptom.

In addition, collateral obstruction, extreme dryness and stasis of yin blood caused by insufficient *ying-wei*, qi, blood, and the insecurity of *wei* qi along with wind-cold attack when healthy qi is insufficient can also lead to inhibition of joint movement. In general, prolonged illness entering the collaterals, obstruction of phlegm-stasis, and external pathogenic attack are a large part of the essential pathomechanism.

As a result, clinical treatment usually emphasizes methods for invigorating the blood and dissolving stasis. For example, points which invigorate blood and move qi, collateral bloodletting therapy, and point injection therapy can be chosen.

Point selection is based on syndrome differentiation and is the key point of treating KOA with acupuncture:

TCM theory states that KOA has both external and internal etiological factors. The internal etiological factors are liver and kidney depletion and deficient yang qi, while the external involve a wind-cold-dampness attack. The disease mechanism is qi and blood obstructing the channels and collaterals. The pattern is root-deficiency with phlegm, and the branch-excess manifest as *bì*. The treatment should adhere to the the rules which include supplementing and boosting the liver and kidney, dispelling wind and dissipating cold, invigorating blood and dissolving stasis, warming yang, and dissolving phlegm. In the clinical setting, knee *bì* should be treated with correct syndrome differentiation based on its manifesting characteristic–root-deficiency and branch-*bì*.

In this case, the principal points include EX-LE4 (*nèi xī yǎn*), ST 35 (*dú bí*), ashi points, ST 34 (*liáng qiū*), SP 10 (*xuè hǎi*), ST 36 (*zú sān lǐ*), GB 34 (*yáng líng quán*); symptomatic point selection includes BL 23 (*shèn shù*), BL 17(*gé shù*), BL 18 (*gān shù*), RN 4 (*guān yuán*), and RN 6 (*qì hǎi*).

Morbidity of the knee joint necessarily influences the normal channel flow around the knee joint, and result in pathological change, channel obstruction and abnormal flow of qi and blood. Where there is stoppage, there is pain. Acupuncture stimulates the local points around the affected area, but distal points also part of the point combinations.

Warm-needling can expand local capillaries and promote blood circulation to absorb inflammation and achieve the effects of warming, dissipating cold, unblocking the channels, quickening the collaterals, dispersing stasis, dissipating masses, and invigorating blood; electronic acupuncture may also diminish inflammation and relieve pain. All of these therapies act to dredge the channels and regulate the qi and blood to improve clinical symptoms according to the principle: "where is free-flow, there is no pain".

Chapter 4

Prognosis

The prognosis varies according to constitution, pathogenic factors, severity, and the treatment plan according to differentiation of signs and symptoms. Generally speaking, the prognosis is improved when appropriate treatment and nursing are provided. When treatment is delayed, patterns of wind *bì*, cold *bì*, damp *bì* and heat *bì* will become prolonged, thus damaging qi and blood and the liver and kidney, resulting in a deficiency of healthy qi and retained pathogens. This may result in a root-deficiency branch-excess pattern which can become an intractable deficiency-excess complex.

Pain and stiffness are two symptoms that first occur in KOA; swelling will appear later on in its development. Pain and swelling have the stages of onset and remission. With the development of KOA, symptoms occur frequently. In later stages, muscles will atrophy, especially the quadriceps femoris, and the knee joint will become deformed. Bow-leggedness is generally present to some degree, and knock-knees may occur in more severe cases. Flexion and extension become inhibited as the range of motion decreases, but completely immovable joints are rarely seen. It should be noted that with appropriate treatment and care in the early stages of KOA, the presenting symptoms can be totally controlled.

Chapter 5

Preventive Healthcare

Dietary control, weight reduction, mineral supplementation including calcium, and proper functional exercise are very important in the treatment of this disease. Additionally, attention to the selection of appropriate shoes, insoles, crutches and kneepads are important. Measures should also be taken to avoid over-strenuous activities that may exacerbate the condition.

Lifestyle Recommendations

Regulation in Daily Life

1. Pay attention to cold and dampness.

Add and/or take off clothes especially when there are changes in the weather. Avoid living in shady, cool and damp places for a long time. Furthermore, avoid exposure to wind when sweating, getting wet in the rain, and taking cold showers after sweating.

2. The living space should be dry with sunlight, bedclothes dry and warm.

Regulation of Diet

Dietary therapy takes advantage of the known properties and flavors of food in order to achieve the functions of dispelling external pathogens, unblocking the collaterals, and eliminating disease, but can also be an important component in the treatment of pain. Dietary therapy must be combined with internal and external therapies, including nursing. While administering dietary therapy, do not underestimate the simultaneous use of other adjunctive therapies.

TCM Food Therapy

Exuberant Wind-Cold-Dampness

1. Noodles with ginger, garlic and chili:

生姜	*shēng jiāng*	10 g	Rhizoma Zingiberis Recens
大蒜	*dà suàn*	10 g	*Bulbus Allii*
辣椒	*là jiāo*	10g	Chili
面条	*miàn tiáo*	100-150 g	Noodles

2. Fáng fēng (Radix Saposhnikoviae) porridge:

防风	*fáng fēng*	10-15 g	Radix Saposhnikoviae
葱白	*cōng bái*	2 stalks	BulbusAllii Fistulosi
粳米	*jīng mǐ*	60 g	Oryza Sativa L.

3. Ròu guì (Cortex Cinnamomi) pulp porridge:

肉桂	*ròu guì*	2-3 g	Cortex Cinnamomi
粳米	*jīng mǐ*	30-60 g	Oryza Sativa L.
红糖	*hóng táng*	To taste	Brown Sugar

Exuberant Heat and Damp Heat

1. Yì yǐ rén (Semen Coicis) porridge:

薏苡仁	*yì yǐ rén*	60 g	Semen Coicis

2. Yì yǐ rén (Semen Coicis) and zhú yè (Herba Lophatheri) porridge:

薏苡仁	*yì yǐ rén*	60 g	Semen Coicis

丝瓜	*sī guā*	100 g	Sponge Gourd
淡竹叶	*dàn zhú yè*	20 g	Herba Lophatheri

3. Dōng guā (white gourd) and yì yǐ rén (Semen Coicis) porridge:

冬瓜	*dōng guā*	500 g	White Gourd
薏苡仁	*yì yǐ rén*	30 g	Semen Coicis

Binding of Phlegm and Stasis

1. Táo rén (Semen Persicae) porridge:

桃仁	*táo rén*	10-15 g	Semen Persicae
粳米	*jīng mǐ*	50-100 g	Oryza Sativa L.

2. Sān qī (Radix et Rhizoma Notoginseng) and dān shēn (Radix et Rhizoma Salviae Miltiorrhizae) porridge:

三七（田七）	*sān qī (tián qī)*	10-15 g	Radix et Rhizoma Notoginseng
丹参	*dān shēn*	15-20 g	Radix et Rhizoma Salviae Miltiorrhizae
粳米	*jīng mǐ*	300 g	Oryza Sativa L.

3. Chuān niú xī (Radix Cyathulae) porridge:

川牛膝	*chuān niú xī*	20 g	Radix Cyathulae
粳米	*jīng mǐ*	120 g	Oryza Sativa L.

Liver and Kidney Deficiency

1. Sāng shèn (Fructus Mori) porridge:

桑椹	*sāng shèn*	20-30 g	Fructus Mori

粳米	*jīng mǐ*	100 g	Oryza Sativa L.

2. Sāng jì shēng (Herba Taxilli), egg and red date porridge:

桑寄生	*sāng jì shēng*	30 g	Herba Taxilli
鸡蛋	*jī dàn*	2 eggs	Egg
大枣	*dà zǎo*	10 dates	Fructus Jujubae

3. Immortal porridge:

制何首乌	*zhì hé shǒu wū*	30-60 g	Radix Polygoni Multiflori Praeparata
大枣	*dà zǎo*	3-5 dates	Fructus Jujubae
粳米	*jīng mǐ*	100 g	Oryza Sativa L.
冰糖	*bīng táng*	To taste	Brown sugar

Yang Deficiency

1. Lù jiǎo jiāo (Colla Cornus Cervi) porridge:

鹿角胶	*lù jiǎo jiāo*	15-20 g	Colla Cornus Cervi
粳米	*jīng mǐ*	100 g	Oryza Sativa L.
生姜	*shēng jiāng*	150 g	Rhizoma Zingiberis Recens

2. Xù duàn (Radix Dipsaci), dù zhòng (Cortex Eucommiae) porridge:

续断	*xù duàn*	30 g	Radix Dipsaci
杜仲	*dù zhòng*	30 g	Cortex Eucommiae
猪尾	*zhū wěi*	1-2 pcs	Pig's tail

3. Gŭ suì bŭ (Rhizoma Drynariae) and mutton porridge:

骨碎补	*gŭ suì bŭ*	60 g	Drynaria rhizome
羊肉	*yáng ròu*	250 g	Mutton

Functional Exercise

Muscular coordinated movement and an increase in energy can relieve pain effectively. In order to increase muscular energy and endurance, ensure to increase mobility of the knee joints. Physical exercise is very beneficial and improves the patient's quality of life. Exercise of the quadriceps femoris muscles definitely have beneficial and obvious effects on relieving pain and improving function.

Exercises for the Quadriceps Femoris Muscle

The patient takes a sitting position and places the upper and lower leg at a 90 degree angle. Hang an object (2-3 kg) on the ankle and contract the quadriceps femoris intentionally to bend and stretch the knee joint. Care should be taken to not lock the knee joint in the extended position, as this may cause injury. Exercise of the quadriceps can increase its power of contraction and promote the flow of blood around the area. It also prevents muscular atrophy and decreases calcium loss.

Exercise on the Bed

With the patient in the sitting position at the edge of the bed, persuade the patient to move the knee and extend the two legs. After this exercise, have the patient bend the knee joint on one side until the sole of the foot touches the bed, then repeat the movement on the other side. This bilateral exercise should be performed alternately for 10-15 repetitions.

Movement of the Toe and Ankle

With the patient in the sitting position, extend the two legs. After that, stretch and bend the great toe and then rotate the ankles internally and externally. Repeat for 10-15 repetitions.

Chapter 6

Clinical Experience of Renowned Acupuncturists

Clinical Experiences of Xie Guo-rong on Treating KOA

Professor Xie Guo-rong treats various arthralgias with the tug-of-war needling technique. He emphasizes the importance of needle manipulation and believes that proficiency in needle manipulation is derived from clinical practice. We should work to standardize the techniques and to form special theories, methods, formulas, point selections and manipulations.

The different types of needle manipulations accumulated throughout the ages are abundant, totally diverse, and unique. It is a known fact that needle manipulation is one of the key points of acupuncture. Unfortunately, most famous doctors inherited their own special needling techniques, and never educated others how to implement them. In addition, needle manipulation is applied by finger force which is difficult, if not impossible, to learn just by watching. Li Shou-xian has stated that the difficulty of acupuncture lies in techniques other than point selection. Needle manipulation is difficult to master, but to cure intractable diseases quickly and effectively in clinic, the doctor must first master the techniques and manipulations.

Treatment Technique - Brief Description

Tug-of-war needling refers to a traditional acupuncture technique where two needles are inserted into two acupuncture points with the two needle tips pointing in opposite directions. Once the needles are inserted, manipulation is initiated. The manipulation specific for KOA is as follows:

The two points chosen are ST 36 (*zú sān lǐ*) and *hè dǐng* (EX-LE2, 鹤顶). After routine disinfection, insert *hè dǐng* obliquely in the posterior-superior direction with a No.30 needle (length: 50 mm) and then insert another needle at ST 36 obliquely in the posterior-superior direction with the same standard size needle. Manipulate the needles until qi arrives. When the qi arrives, the operator will feel tightness around the needle, just like the sensation of a "fish taking the bait", and the patient will feel soreness, numbness, distension, heaviness and heat when the qi arrives.

Next, the operator holds the two needles handles with both hands and presses the needles with finger force to make the two needle tips oppose each other. The finger force is required to pass through the needle tips which cause the needling sensation (soreness, distension and heat) to radiate, connect and spread out to the area around both acupuncture points.

Next, the operator holds the needle handles and lifts the needles with opposing force, like pulling the needles out, but does not withdraw the needles. Two seconds later, repeat the step of inserting the needles with finger force. In this way, the two needles are alternately pressed and lifted as if a rope is being pulled and dragged inside the knee joint in order to transmit the needling sensation up and down, between the two points. This is why the technique is called "tug-of-war" needling. After three minutes, the patient will feel the sensation of heat inside the knee joint. The last step is to withdraw the needle and press the needle hole with a disinfected dry cotton ball.

Regarding the patient that obtains no obvious qi sensation, the operator can repeat the method for another three minutes. Do not continue with this technique if the qi sensation is not perceived by the patient. If no qi sensation is obtained, the operator should just remove the needles and apply a different needling technique.

The key points of the tug-of-war needling technique are the opposing needle tips and the rule of "when the qi arrives, there is therapeutic efficacy". It induces the radiation of qi sensation through pressing the needle body and produces a heat sensation in the knee joint. This method has the characteristic of achieving rapid effect and produces a comfortable sensation for the patient. Currently, the tug-of-war needling techniques mechanism of pain relief is not fully understood, and further research is required.

Case Study

Male, age 45

First Visit: February 23th, 2003

【Chief Complaint】

Bilateral knee joint pain and lack of strength for 3 years

【Medical History】

Three years ago the patient was performing heavy labor. Afterward, the patient suffered from pain in both knee joints which was gradually aggravated by overstraining and changes in the weather. The pain was accompanied by a lack of strength when walking, and inhibited range of motion when bending and stretching. After the illness, his appetite was poor, and although he was sleeping well, his sleep was disturbed by urinary frequency, urinating as many as 2-3 times each night. His defecation was normal. His tongue was pale with teeth marks and the coating was a little bit greasy. His pulse was thready.

【Examination】

The patient was obese, appeared fatigued, and his knee joints were swollen and deformed. There was tenderness all around both knee joints. Movement was inhibited when attempting to assume a squatting position, or attempting to stand after prolonged sitting. There was a sensation of friction when the knee joint was mobilized but the color of the skin around the knees was not red and its temperature was low when palpated.

X-ray films revealed lip-like hyperosteogeny on the articular surface.

【Diagnosis】

TCM diagnosis: knee *bì* (qi deficiency with exuberant dampness)

Western diagnosis: Knee Osteoarthritis (KOA)

【Pattern Differentiation】

The intensive labor and obesity produced an over-loading of the knee joints. The overstrain damaged the articular cartilage, causing the articular space to narrow, causing hyperosteogeny. Additionally, the patient is over fifty years old and his kidney qi had gradually declined. Since the kidney governs the bone, this aggravates the pain and causes the lack of strength. Insufficient kidney qi fails to astringe, so the symptoms such as frequent urination and nocturia manifest. Deficient qi fails to dissolve the water-dampness causing the middle *jiao* to become obstructed, causing his poor appetite. The pale tongue with a white greasy coat and teeth marks are also signs of kidney qi depletion and water-dampness obstructing the middle *jiao*.

【Treatment Principles】

The treatment method consists of dispelling dampness and unblocking the collaterals, supplementing the kidney, and strengthening the bone. The needling method utilized will be the tug-of-war technique, using even manipulation.

【Treatment】

Point Selection

ST 36 (*zú sān lǐ*)	EX-LE2 (*hè dǐng*)

Manipulations

Doctor Xie selected *hè dǐng* on the upper position, and ST 36 on the lower position of the painful area. In order to transmit the needling sensation to the diseased area, the needle tip was placed facing toward the direction of the illness. When the qi arrived, the operator utilized the twirling method for 1-2 minutes causing the channel qi under the tip of the needle to radiate and spread around.

Afterward, the operator held and pressed the handles of the two needles for 1 minute in order to ensure the needling sensation under the two hands connected and aligned with each other, at which time he either withdrew the two needles or retained them for 10-20 minutes. When the qi arrived, the patient felt the needle sensation flow up and down, generating a heat sensation in the knee joint, resulting in obvious relief of his knee pain.

【Follow-up】

When the patient consulted on the next day, he stated that his pain was alleviated, that range of motion was improved, and the tenderness was relieved. This provided evidence that the syndrome differentiation and treatment were effective, so the tug-of-war needling technique was continued as before.

The operator selected *hè dǐng* and ST 36 bilaterally. When qi arrived, the operator applied the twirling method for 1-2 minutes to make the channel qi under the tip of the needle to radiate and spread. After that, the operator held and pressed the handles of the two needles in order to ensure the needling sensation connected and aligned. He then retained the needles for 10-20 minutes, and then withdrew the two needles.

On the third treatment, the pain was relieved and the operator continued with the tug-of-war needling technique bilaterally on *hè dǐng* and ST 36, additionally, he included filiform needling on BL 40 (*wěi zhōng*), RN 6 (*qì hǎi*) and SP 10 (*xuè hǎi*) as auxiliary therapy.

After six continuous treatments, the patient stated that his stubborn pain disappeared; the range of motion improved, being flexible and easy when walking, squatting and bending. His tongue was light red with a thin coating, and the pulse was normal. The patient appeared well and was advised to concentrate on keeping warm and avoiding long journeys. After a three month follow-up, the illness did not relapse.

【Discussion】

The patient is over fifty years old; his overstrain and decline in kidney qi led to pain in the knee joints and lack of strength when walking. Insufficient kidney essence and a weak constitution aggravated the pain by overworking and weather changes.

Comprehensive Medicine According to Master Zhang (*Zhāng Shì Yī Tōng*, 张氏医通) states: "the knee is the residence of sinew……knee pain is invariably attributed to deficiency of the liver and kidney. Deficient conditions always tend to be susceptible to attack by wind, cold and dampness".

The Yellow Emperor's Inner Classic: Basic Questions states: "impairments with five kinds of strain, prolonged standing impairs the bone, prolonged walking impairs the sinew" and deficient qi fails to transform water-dampness.

As a result, the water-dampness obstructs the middle *jiao* and affects the movement of qi, so the appetite becomes poor. A light pale tongue with a white greasy coat and teeth marks are also signs of depletion of kidney qi and water-dampness obstructing the middle *jiao*.

The Yellow Emperor's Inner Classic: The Spiritual Pivot states to "retain the needles for cold conditions"; the therapeutic principles are to dispel dampness, unblock the collaterals, supplement the kidney and strengthen bone. Needling ST 36 and *hè dǐng* complies with the rule of "the indication extends to where the channel reaches".

Selection of local points can dredge the channel and directly alleviate the pain. ST 36 can fortify the spleen and stomach, dissolve dampness in the middle jiao and assist digestion. Appropriate therapy can supplement healthy qi and regulate the yin yang of the *zang-fu* organs so the pathogen can be dispelled and the pain relieved.

Dr. Xie has attained great achievements in needling methods for the arrival of qi. Just as *The Spiritual Pivot - Treatise on Nine Needles and Twelve Channels* (*Líng Shū – Jǐu Zhēn Shí Èr Yuán*, 灵枢•九针十二原) states: "the rule of acupuncture is that when the qi arrives, there is therapeutic efficacy".

Dr. Xie believes that to obtain an obvious effect, the arrival of qi at the location of the disease is more important than the arrival of qi in general. In his decades of clinical practice, Dr. Xie has explored several methods to promote the arrival of qi at the location of the disease. Some of these techniques follow below:

1. The Pressing Method:

In this technique, the needle tip points toward the affected location and the needle is manipulated to promote the arrival of qi. Then hold the needle handle and concentrate qi to the needle tip and press the needle with finger force to stimulate the arrival of qi at the location of the disease. Proper application can produce rapid effects. Some feel that it produces a sensation of heat at the affected area.

After the arrival of qi at the affected area, the pain is immediately relieved.

2. Tug-of-War Needling Technique:

Select two points, one above and one below the affected area. The needle tips point toward the diseased or injured region. After the arrival of qi, exert slight and even twirling manipulation for 1-2 minutes. Then hold the needle handles and press with finger force for over 1 minute until the needling sensation radiates aligns and connects. Finally, either withdraw or retain the two needles. This method is frequently used on *bi* syndrome with a limited scope, like *bi* syndrome of the knee joint area. We can select *hè dǐng* on the upper position and ST 35 or ST 36 on the lower position. Apply this method until the needling sensation radiates, connects and aligns. It will produce a sensation of heat around the affected area and the pain will immediately be relieved.

3. Relay Needling:

It is also called "continuous racing-horse needle". Select two points whose distance is not too far apart. Exert more intensive stimulation on the lower point than on the upper point. Do not withdraw the upper needle until the swelling sensation disappears. Then insert another needle under the lower point, which forms a new couple, and apply needling manipulation like before. Repeat this method, continuously moving downward like a relay race, until the last needle reaches the end of the affected limb. This needling method is indicated for lumbocrural pain, periodic paralysis, and other specific conditions.

The points mentioned below are frequently selected: DU 3 (*yāo yáng guān*), BL 25 (*dà cháng shù*), GB 30 (*huán tiào*), BL 40 (*wěi zhōng*), GB 34 (*yáng líng quán*), BL 60 (*kūn lún*), GB 40 (*qiū xū*), GB 41 (*zú lín qì*) or KI 1 (*yǒng quán*).

Clinical Experiences of Dong Jing-chang on Treating KOA

Dr. Dong Jing-chang's ancestral homeland is Ping Du County, Shandong

Province, where he inherited and accepted his family's medical lineage when he was a child. When he was 18, he practiced as a doctor in order to help people, and gained a widespread reputation. In 1949, his whole family moved to Taiwan. He has diagnosed over 400 thousand patients and has saved numerous lives over the decades. In 1971, Doctor Dong used extra points to cure hemiplegia in the president of Khmer, Lon Nol. The acupuncture cure was viewed as miraculous and resulted in a sense of shock all over Taiwan. As a result, Doctor Dong gained the high reputation of a "contemporary acupuncture sage". In 1973, Doctor Dong broke the ancestral protocol to never introduce their medical theory and practice to people with a different surname and started to impart his knowledge to his apprentices. He then edited and published *Dong's Acupuncture Channels and Extra Points* (*Dōng Shì Zhēn Jiǔ Zhèng Jīng Qí Xùe Xúe*, 董氏针灸正经奇穴学). From then on, this unique ancestral medical lineage which was held secret for over ten generations was spread throughout the world.

"Dong's extra points" is a unique ancestral achievement and is different from the fourteen regular channel systems. Dong's system consists of unique extra points, needling manipulations, and diagnosis methods all with strong clinical practicality.

Brief Description

Dong's extra points have their own special "collateral theory", which holds that chronic disease must result in stasis, abnormal disease must lead to stasis, severe disease must result in stasis, and pain must also bring about stasis. Any case that cannot be alleviated after several acupuncture treatments must have blood stasis present. Dong believes that blood stasis must obstruct the movement of qi, so we should find the static collateral and use the bloodletting technique to expel the pathogenic qi and blood. With this treatment method, many severe lingering and obstinate diseases can be cured.

1. The Function of Collateral Bloodletting Therapy:

This technique removes coagulation, opens stagnation, dispels stasis, drains heat, invigorates blood circulation, expels toxins, and drains dampness.

2. The Characteristic of Dong's Bloodletting Therapy:

(1) Bleeding the remote area which corresponds to the ancient doctrine, "take collateral bloodletting therapy with remote needling".

(2) The needling locations are all over the body, which can be divided into many areas, such as the heart and lung area, the liver and gallbladder area, the kidney area and so on.

(3) The therapeutic range is wide, the curative effect is rapid and the needling manipulation is easy and safe. Dong's bloodletting therapy is a perfect combination of filiform and three-edged needling, which conforms to the two statements: "arrival of qi at the affected area" and "pathogens have a way out". It also reflects the true integration of body and mind, and the therapeutic principle of dual regulation of the body and spirit.

Case Study

Female, age 58

First Visit: May 10th, 2000

【Chief Complaint】

Patient reports pain of the right knee joint for 5 years with gradual aggravation over the past half month.

【Medical History】

The patient always stands when at work, which is very strenuous and causes her to have difficulty squatting; inhibited bending and stretching during movement sometimes causes her to stumble. Therefore, her work and life have been seriously affected. She has taken medicinal tablets and herbs, but her pain has not been alleviated. She applied a plaster for external application, but her skin was seriously allergic to the plaster, so she stopped using it.

【Examination】

The two knee joints are swollen and deformed with tenderness all around the affected areas. The bending and stretching movement is inhibited. The squatting movement is also inhibited, and standing up is difficult after squatting or sitting for a prolonged duration. There is a sensation of friction when moving the knee joint. The tongue is dark with a thin and white coat, and the pulse is deep and choppy.

X-ray film revealed the knee joints suffered from hyperosteogeny and the cavity has become more narrow than normal.

【Diagnosis】

TCM diagnosis: knee *bì* (internal obstruction of blood stasis, liver and kidney deficiency)

Western diagnosis: Knee Osteoarthritis (KOA)

【Pattern Differentiation】

Long term standing causes overloading of the knee joints, so the overstrain results in articular cartilage damage, narrowing of the joint-space and hyperosteogeny. The patient is over fifty years old, and the condition of the liver and kidney are declining and becoming deficient. Deficiency of the liver and kidney leads to weakness of the bone and sinews. Liver and kidney deficiency cause inhibited walking because the kidney governs the bone and the liver governs the sinew. Many therapies had no effect, because the patient had suffered from the disease for such a long time. Long-standing cases must have stasis; abnormal cases must have stasis, and thus the patient a dark tongue and a deep choppy pulse.

【Treatment Principles】

Moving stasis and unblocking the collaterals while supplementing and nourishing the liver and kidney. The needling technique used will be the even method, followed by cupping and collateral bloodletting.

【Treatment】

Point Selection

Points: ST 35 (*dú bí*), GB 34 *(yáng líng quán)*, SP 9 (*yīn líng quán*), *hè dǐng* (EX-LE2) and *sān jīn* (Dong's Three Extra Points: The three points are level with the spinous processes of the 3rd , 4th, and 5th thoracic vertebrae, 3 cun lateral to the posterior midline, on the second line of foot taiyang bladder channel).

Manipulations

Needle ST 35, GB 34, SP 9, *hè dǐng* on the affected limb; apply electro-acupuncture with an intermittent wave. Retain the needles for 20 minutes; then needle *sān jīn*, Dong's extra point on the back.

Cupping following collateral bloodletting therapy is given one time per day; 10 days form one therapeutic course of treatment. After two courses, the painful and swollen knee joint is obviously relieved. The patient can squat slowly and her walking is more powerful than before. After two courses of treatment, the illness is cured completely.

【Discussion】

Dong believes that chronic disease must result in stasis, abnormal disease must lead to stasis, severe disease must result in stasis, and pain must bring about stasis. The patient has suffered from the disease for a long time, so her kidney qi has declined due to overstrain. Deficient kidney qi can lead to pain and lack of strength because the kidney governs the bone. In addition, kidney essence declines and the constitution becomes weak when people age, especially older women, for this reason her disease was not alleviated after several treatments of acupuncture therapy.

In this condition, Dong believes that blood stasis must obstruct the movement of qi, so we should find the static collateral in the affected area and perform collateral bloodletting to expel the pathogenic qi and blood.

The Yellow Emperor's Inner Classic states: "what humans possess are blood and qi". Primary morbidity often lies in the channel because the channel governs qi; prolonged disease often attacks the collateral because the collateral governs the blood. Any lingering or obstinate disease that suffers from the pathologic course of qi stagnation, blood stasis, phlegm coagulation or the gathering of toxins, often results in stasis as the primary turning point of the disease.

Dong uses filiform needles to unblock the channels and regulate the qi, and employs three-edged needles on the collaterals to invigorate blood. Afterwards, qi moves without blockage, blood flows smoothly, and there is no need to worry about the radical cure of the disease. Dong's extra points give a vivid respect in regard to the essence of acupuncture. Only by great concentration and repetitive practice can the practitioner comprehend this clearly.

Chapter 7

Perspectives of Integrative Medicine

Challenges and Solutions

KOA refers to a disease that is marked by degeneration of articular cartilage which involves the whole joint, including cartilage, ligaments, capsule, synovium and the musculature around the joint. Its pathomechanism is not completely clear, but it is not difficult to diagnosis. Although it is easy to temporarily relieve the symptoms through medical treatment, it is very difficult to stop and completely cure the disease and to prevent relapse. Terminal KOA is marked by severe damage to the articular bone, cartilage, ligament, capsule and synovium which cannot be cured effectively by non-operative therapies.

Preventing Occurrence and Relapse

Although the pathomechanism is not completely clear, the holistic concept of TCM plays an important role in guiding the treatment and prevention of KOA. The internal cause of KOA is deficient essential qi of the liver and kidney failing to nourish the sinew and bone. TCM theory states that the liver governs sinew and the kidney governs bone, while the external factors include overstrain, traumatic injury and attack of pathogenic wind, cold and dampness.

As for prevention, avoid traumatic injury and overloading of the joints. Other beneficial factors include weight loss, knee joint protection in daily life, and the avoidance of wind, cold and dampness. Additionally, proper physical exercise can maintain the flow of qi and blood, and keep the knees in good condition.

From a therapeutic standpoint, people should take some herbal prescriptions and specially prepared foods for supplementing and boosting the liver and kidney, such as *Liù Wèi Dì Huáng Wán* (六味地黄丸, Six Ingredients Rehmannia Pill), *gǒu qǐ zǐ* (枸杞子, Fructus Lycii), stewed walnut and rabbit meat. These foods and medicines are recommended because as people approach middle age and *tiān guǐ* gradually declines, the essential qi of the liver and kidney become insufficient.

Moreover, emphasis should be placed on adjustment of daily life, such as maintaining a healthy diet and living habits, avoiding alcohol damage, and excessive sexual intercourse. All of the measures mentioned above have a significant effect on preventing KOA.

Curing Severe Cases of KOA

KOA symptoms include redness, swelling, hot and painful knee joints, and inhibited range of motion. These symptoms are frequently seen in the elderly, especially in women after menopause. Western medicine often relies on treatments such as reduced movement and rest, NSAIDs, local block therapy with hormones and physical therapy, but the effects are not satisfactory.

TCM considers knee *bi* as a pattern of root-deficiency and branch-excess. The treatment should follow the principle of "treating secondary symptoms for emergency, while treatment of chronic disease aiming at the principal cause" With acute onset in the elderly, the "branch" is the obstruction of wind, cold, dampness, heat, phlegm and stasis, while the "root" is deficient liver and kidney, and insufficient qi and blood. So the therapeutic principles of treating the "branch" include dispelling wind and dampness, clearing heat, dispelling stasis, unblocking the channels, relieving pain, and temporarily stopping movement. After the acute stage is over, the emphasis should be placed on treating the "root".

Generally, principles such as enriching and nourishing the liver and kidney, relaxing the sinews and quickening the collaterals is used, or warming the kidney, strengthening the yang, warming the channels and unblocking the collaterals, while at the same time increasing the movement of the knee joints.

In later stages, KOA needs operative treatment, because conservative therapy is not effective and KOA occurs repetitively and influences daily life seriously. Operative treatment aims at improving physical structure, relieving pain, improving joint function and increasing the quality of life. Its immediate convenience can't be replaced by medicines. Advanced equipment lays the foundation of operative techniques and its efficiency.

However, there is a lot of medical literature reporting the possibility that an operation may even induce KOA. In one research study, Moskowitz and his colleagues confirmed through animal experimentation that meniscus excision induced KOA. As a result, some scholars suggest that the doctor should consider all practical approaches, carefully before an operation and inform the patient of the risks and benefits of each approach.

Acupotomy is a recently invented surgical therapy that can go right to the affected area and effectively dissect the adhesion and spasm. Improving the tissue's deformity and degeneration and restoring the articular mechanics back to their normal condition. Although its mechanism of action needs further research and explanation, acupotomy can also relieve clinical symptoms through self-regeneration and restoration of the knee joint. Effective treatment on soft tissue can be applied at every stage of KOA, so acupotomy can be viewed as the possible preferred therapy of the future.

Insights from Empirical Wisdom

The onset of KOA progresses into a chronic and progressive disease process, it is acute and remission stages alternate. The therapeutic objective is pain relief and maintaining free movement.

The key points include three aspects: pharmacologic therapy, combination of movement and rest and functional exercise. Functional exercise and pharmacologic therapy as well as internal and external medicine, are all equally important. The so-called combination of movement and rest refers to proper rest in the acute stage and planned and intentional exercise in the remission stage. Pharmacologic therapy should comply with *bi*'s fundamental characteristic of "root-deficiency and branch-excess". During the acute stage, treating *bi* should be emphasized, while during the remission stage, performing physical therapy and appropriate rest is helpful to decrease the potential for relapse.

The pathomechanism of KOA is not clear, some TCM therapies and Western medicines have certain effects, but all of them fail to cure the disease completely; repetitive occurrence is one of its characteristics. Western therapies such as NSAIDs and local block therapy with hormones can provide short-term effect, but their side-effects are very serious; for example, NSAIDs stimulates the gastrointestinal system seriously, hormone abuse induces endocrine disorders such as bone necrosis, osteoporosis and relapse. TCM therapies have fewer negative side-effects, better efficacy and are easier to manage. Moreover, after the acute stage, the therapeutic principles of TCM change into supplementing and boosting the liver and kidney, strengthening the bones and muscles, relaxing the tendons and activating the channels. As long as the therapies above are followed strictly, efficacy can be further ensured.

In addition, TCM places emphasis on functional exercise. TCM believes that people with different occupations should perform different exercises, for example, the white collar worker and sedentary people should do more exercises such as tai chi, jogging, bicycling, and swimming. Laborers or people with an overabundance of physical activity in daily life should have massages, keep warm, and ensure occupational protection.

Summary

Recently, the incidence of KOA has been gradually increasing. TCM has its own specialties and advantages. Because the etiology and pathomechanism of KOA is not totally understood, the therapeutic method of western medicine continues treating the branch which is represented by the conventional pyramid model whose foundation includes education, physical therapy, occupational therapy, weight loss, physical exercise and auxiliary devices.

The patient who fails to respond to the therapies mentioned above should administer analgesics such as Panadol; if there is still no response, consider the administration of NSAIDs; the case that has no effect after internal treatment, should consider surgical therapy. So far, there is no news reporting a medicine whose action directly treats KOA.

From the research performed by different experts, we can conclude that acupuncture and TCM can improve the function of cartilage cells, promote cartilage repair, inhibit the synovial inflammation, postpone cartilage degeneration, improve microcirculation and hemorrheology, and decrease intraosseous pressure.

With the development of modern medicine and the integration of Chinese and Western medicines, experimental acupuncture will inevitably provide a great leap forward in development. In light of the problems at present, it can be researched as follows:

Literature Review

We should review all the ancient writings concerning KOA and diseases with similar symptoms as completely as possible. After that we should generate and arrange the property, degree and accompanying symptoms. Then analyze the KOA's related symptoms and illnesses, and determine the typical ones as the foundation of syndrome differentiation. Finally, we should discuss KOA and the diseases similar to KOA in the context of TCM and seek out their similarities and differences as a therapeutic reference.

Survey of Epidemiology

We should investigate the condition, looking at the incidences of sex, age, occupation, and education by means of a large scale sample study.

Clinical Research

Generally and systematically investigate TCM's symptoms on the patients suffering from KOA, then seek out the common symptoms and establish a complete syndrome differentiation system of TCM through analysis and summary.

Standardized Evaluation

Make a classified observation on the basis of disease evolution and evaluation of the curative effect, and continue with a follow-up survey. Carry out random, double-blind, contra positive research and evaluation test including large samples with a multicenter. Set down a unified standard of diagnosis and then quantify the therapeutic standard of TCM's symptoms and treatment; in addition, standardize the observation indicators and therapeutic effects. In terms of TCM's symptoms, establish a standard diagnosis and treatment which comply with Chinese medical theory under the reference of standard diagnosis and treatment made by the ARA.

KOA has a very complicated etiology, but it is confirmed that biomechanical factors have an important effect. So the treatment should be emphasized on holistic therapy by which the advantages of TCM can be applied. In a clinical setting, in order to obtain a better effect of pain relief and functional improvement, it is necessary to enrich TCM's external therapies. Besides conventional acupuncture and the application of compresses, intra-articular therapies such as medicinal lavation under articularscope, intra-articular injection and medicinal iontophoresis, can have a long term action on the intra-articular affected area, which is significant in the clinical treatment and evaluation.

Chapter 8

Selected Quotes from Classical Texts

1.《素问·痹论》:“风寒湿三气杂至,合而为痹也。其风气胜者为行痹,寒气胜者为痛痹,湿气胜者为着痹也”

“The three types of pathogenic qi which are wind, cold and dampness combine and result in *bì* syndrome in which the predominance of wind leads to migratory (wandering) *bì*, the predominance of cold leads to painful *bì* and the predominance of dampness leads to fixed *bì*.”

Basic Questions - Discussion on bì (*Sù Wèn – Bì Lùn*, 素问•痹论)

Explanation: *Bì* syndrome is caused by the inhibited flow of qi and blood due to exogenous wind-cold-dampness and attack of the channels. The predominance of wind is called migratory *bì*, the predominance of cold is called painful *bì* and the predominance of dampness is called fixed *bì*.

2.《素问·举痛论篇》:“寒气客于脉中则气不通”

“Qi is blocked when cold invades the channel.”

Basic Questions - Discussion on Pain (*Sù Wèn - Jŭ Tòng Lùn*, 素问•举痛论)

Explanation: Cold can lead to vessel contraction which slows the flow of qi and blood and causes vessel obstruction, because cold dominates contraction.

3.《灵枢·寿夭刚柔》:“久痹不去身者,视其血络,尽出其血”

“When the patient suffers from a prolonged disease, the vessels should be observed and needled in order to expel extravasated blood”

The Spiritual Pivot - Treatise on Longevity and Dying Young, Strong and Soft (*Líng Shū – Shòu Yāo Gāng Róu*, 灵枢•寿夭刚柔)

Explanation: Prolonged *bì* syndrome is often accompanied with blood stasis.

Blood stasis can be cured by bloodletting therapy on the raised tiny blood vessels on the surface of the skin.

4.《玉龙歌》:"膝头红肿不能行,必针膝眼膝关穴,功效须臾病不生"

"A red and swollen knee joint with inhibited movement must be treated by needling *xī yǎn* (EX-LE 5, 膝眼). The therapeutic effect is rapid and the symptoms never relapse"

The Song of Jade Dragon (*Yù Lóng Gē*, 玉龙歌), by Ming Dynasty Physician Yang Ji-zhou.

Explanation: A red swollen knee joint accompanied with difficulty walking can be treated with acupuncture needling EX-LE5 (*xī yǎn*) with a rapid beneficial effect.

5.《针灸大全·卷一·治病十一证歌》:"肘膝疼时刺曲池,进针一寸是便宜。左病针右右针左,依此三分泻气奇。膝痛三分针犊鼻,三里阴交要七分"

"A painful elbow and knee can be needled on LI 11 at a depth of 1 *cun*. If the right side is affected, needling is performed on the left side, while if the left side is affected needling is performed on the right side. Drain the qi at a depth of 1/3 of 1 *cun*. A painful knee joint can be needled using ST 36 at a depth of 1/3 of 1 *cun* and acupuncture manipulation applied seven times."

The Complete Compendium of Acupuncture and Moxibustion – Volume one – Treatment Principles for Eleven Syndromes (*Zhēn Jiǔ Dà Quán – juàn yī – Zhì Bìng Shí Yī Zhèng Gē*, 针灸大全•卷一•治病十一证歌), by Ming Dynasty Physician Xu Feng.

Explanation: It is appropriate to needle LI 11 with a 1 *cun* needle when treating a patient with a painful elbow and knee. If the affected area is on the left side, the acupoint on the right side should be needled, and vice versa. The wonderful efficacy is easily obtained when draining manipulation is applied at the depth of 1/3 of 1 *cun*. For example, ST 35 is often needled at a depth of 1/3 of 1 *cun* to treat a painful knee joint. Additionally, acupuncture manipulation is often applied at ST 36 and SP 6 for a total of seven times.

Chapter 9

Modern Research

KOA is usually caused by various factors, either through the same or multiple mechanisms such as mechanics, chemicals, biology, enzymes or other factors, but its specific etiology is still not very clear. Each has the same or similar symptoms whose major manifestation is joint dysfunction. The main reason is articular cartilage degeneration, resulting from continuous and constant action, overloading, and unbalanced stress. Some studies reveal that KOA is related to trauma, inflammation, senility, metabolism and immunity, but again, at present its exact mechanism is not clear. Currently, the epidemiology makes it clear that KOA has some risk factors such as age, sex, obesity and other racial genetic factors. For the past few years, the studies have mainly focused on the pathological degeneration of the articular cartilage which has been confirmed as the fundamental cause, with its pathological essence being an internal and inflammatory disease. Almost all of the factors affect the knee joint and its cartilage, causing dysfunction of qi and blood circulation which is another crucial step in the disease process.

Currently, acupuncture varies from point selection and combination to needling techniques. Acupuncture, combined with point injection therapy, cupping, *tuina*, moxibustion, medicinals and other modern physical medical devices has made definite progress[1].

Point Selection and Combinations

Single Point Selection

Li Xiao-hao[2] treats the disease with a single needle, LI 11 (*qū chí*), and advises the patient to move the affected leg simultaneously and then withdraws the needle after the symptoms are relieved. LI 11 is *he-sea* point of foot *yangming* stomach channel which has the functions of invigorating blood circulation, relieving pain, unblocking the channel and invigorating the collateral; this acupuncture point combined with movement of the affected leg is for the purpose of inducing needling sensation and promoting the flow of qi and blood in the affected area, since the normal flow of qi and blood can relieve pain.

Deng Bai-ying[3] selects PC 6 to treat KOA. PC 6 pertains to the hand *jueyin* pericardium channel. It can dredge the channel qi of *jueyin* and *taiyin*. Qi invigorates the flow of blood leading to the relief of pain.

Local Point Selection

Wang Shu-qin[4] treats the disease by means of obliquely inserting GB 34 (*yáng líng quán*), SP 9 (*yīn líng quán*), SP 10 (*xuè hăi*), ST 34 (*liáng qiū*) with the even needling method because stimulating the points around the knee joint can regulate the channel qi, dredge the vessels, and relieve pain.

Distal Point Selection

Zhu Qing-jun[5] inserts RN 12 (*zhōng wăn*), RN 4 (*guān yuán*), ST 26 (*wài líng*), SP 15 (*dà héng*) combined with ST 24 (*huá ròu mén*) and extra points with abdominal needles. The four principal points are used to regulate the qi and blood and dredge the channel qi, transporting it to the extremities of the four limbs.

Han Yue-dong[6] stimulates the zone of movement and feelings of the foot (足运感区) with the frequency of 200 cycles/minute, and then inserts SP 6 (*sān yīn jiāo*), KI 3 (*tài xī*) and *ashi* points with electricity after the qi arrives. Stimulating this zone can unblock the channel and invigorate the collaterals; SP 6 and KI 3 can strengthen the healthy qi to eliminate pathogens; the *ashi* points can dredge the sinew and unblock the collaterals; electro-acupuncture can enhance the functions while subduing inflammation and relieve pain.

Multiple Needling Manipulations

The needle has developed from the basic filiform and fire needle to electro-acupuncture, needle-shaped knives and microwave needles, which are new instruments combined with modern medical achievements. The penetration method, auricular and abdominal acupuncture are frequently used, and other methods such as triple needling, straight and side needling, encircling acupuncture, great needling, superficial needling and subcutaneous acupuncture are also reported.

Filiform Needling

Chen Xiang-fei[7] selects ST 36 and *hè dǐng* as the principal points. First he inserts supplemental points, and then the principal ones; ST 36 is inserted before *hè dǐng*. The needle tip is inserted rapidly and upward with even manipulation in order to induce and connect the needling sensation to knee joint.

Li Feng[8] treats KOA with the valley union needling technique in which ST 35 (*dú bí*), ST 33 (*yīn shì*), SP 10, ST 32 (*fú tù*), ST 31 (*bì guān*) and other effective points are selected.

Qiu Ling[9] applies the encircling acupuncture technique with 4-8 needles inserted around the tender areas of the affected leg, or on the swelling knee joint, assisted with massage.

Acupoint Pressing Massage

Hong Shu-chen[10] employs acupoint pressure method on the points around the knee area.

Jin Jian-ming[11] applies the rolling method on the quadriceps muscle of the affected leg, especially the upper area of knee joint, and presses *hè dǐng*, ST 34, SP 10, and ST 32; alternately using pressing, kneading, flicking and plucking methods on the ligaments of knee joint and the patella, medial collateral ligament and lateral collateral ligament, and then presses ST 35, *hè dǐng* and GB 34, after that he grasps the patella; using the rolling method on the posterior side of the thigh, popliteal fossa and the posterior side of the shank, and then presses BL 40 (*wěi zhōng*) and BL 57 (*chéng shān*), BL 40 should be treated prior to BL 57; application of the rolling method should be performed with the knee joint in the bent position, combined with passive movements such as bending, stretching and rotation of the knee joint inward and outward. In the final phase of treatment, use the scrubbing method until the area is heated, which assists with muscle strength training.

Moxibustion

Yang Zhi-qin[12] treats this disease with moxibustion on ST 36 and also with electronic needling.

Sun Kui[13] reported that he has always treated the pattern of liver and kidney deficiency with moxibustion using a *fù zǐ* cake.

Electro-Acupuncture

Zhang Guang-li[14] inserts *nèi xī yǎn*, ST 35, GB 34, SP 9, ST 36 and GB 33 (*xī yáng guān*) with filiform needles. After the qi arrives the needles are connected to an electro-acupuncture device using a sparse-dense wave form. He also uses a TDP lamp simultaneously.

Wu Ji-sheng[15] inserts *nèi xī yǎn*, ST 35, GB 34, SP 9, SP 10 and ST 34 supplemented by BL 40, BL 39 (*wěi yáng*) and BL 57. After the qi arrives, the needles are connected with an electro-acupuncture device using an intermittent or a sparse-dense wave form.

Point Injection Therapy

Wang Zong-jiang[16] alternately injects *angelica* and Xylocaine into ST 35, *nèi xī yǎn*, GB 34, SP 9, ST 36 and any applicable ashi points of the two knee joints.

He Cheng-qi[17] injects danshen root into GB 34, and SP 9 and applies movement of the knee joint as a supplemental adjunctive therapy.

Wang Yun-xia[18] uses point injection therapy and medicinal compress therapy to treat the disease. The injections include *dāng guī* (Radix Angelicae

Sinensis), *hóng huā* (Flos Carthami), and *bái sháo* (Radix Paeoniae Alba). The points include ST 36, GB 34, SP 9, SP 10, and BL 23 (*shèn shù*).

Fire Needling

Li Ya-dong[19] treats the disease with the fire needle method on *ashi* points supplemented with ST 36, GB 34, SP 9, SP 10 and ST 35. Other points without fire needle can be stimulated by filiform needle with even manipulation. To the patient with articular hydrops (swelling from excessive accumulation of watery fluid in cells, tissues, or serous cavities), he orders the patient to strain the knee joint and inserts ST 35 with a thin fire needle. After that, he squeezes the knee joint to empty the blood and fluid. He then presses the needle hole with dry sterile cotton ball.

Dong Gui-fang[20] uses the fire needle combined with medicinal compress application to treat KOA. First, the patient bends the knee joint slightly; the doctor then searches for the *ashi* point, and he only treats one point for each treatment. After that, the doctor disinfects the tender spot and inserts the fire needle rapidly. Then withdraws the needle and presses the needle hole instantaneously. For the last procedure, he applies a medicinal compress which include *liú jì nú* (Herba Artemisiae Anomalae), *huā jiāo* (Pericarpium Zanthoxyli), *chuān wū* (Radix Aconiti), *rǔ xiāng* (Olibanum), *mò yào* (Myrrha), *bái jiè zǐ* (Semen Sinapis), *ròu guì* (Cortex Cinnamomi), *zhāng nǎo* (Camphora) and *bīng piàn* (Borneolum Syntheticum).

Acupotomology Therapy

Song Gao-feng[21] utilizes acupotomy therapy and applies *tiān nán xīng* (Rhizoma Arisaematis), *chuān wū* (Radix Aconiti), *cǎo wū* (Radix Aconiti Kusnezoffii), *wēi líng xiān* (Radix et Rhizoma Clematidis), *tòu gǔ cǎo* (Herba Vaccinii Urophylli), *jiāng huáng* (Rhizoma Curcumae Longae), *xì xīn* (Radix et Rhizoma Asari), *hóng huā* (Flos Carthami), *rǔ xiāng* (Olibanum), *mò yào* (Myrrha), *dāng guī* (Radix Angelicae Sinensis), *bái jiè zǐ* (Semen Sinapis), *niú xī* (Radix Achyranthis Bidentatae) and *gǔ suì bǔ* (Rhizoma Drynariae).

He Tie-hao uses acupotomy therapy to treat this disease. The specific steps are as follows: find the exact tender spot according to the patient's X-ray film. After routine disinfection, insert the needle until the patient feels soreness and distention. Then exert the longitudinal and transverse stripping and debonding method one time, and withdraws the needle. This therapy can be accompanied by application of medicinal compress therapy.

Li Jian-dong[22] emphasizes the treatment of the patello-femoral joint. Take the lateral side of the patello-femoral joint as an example: first select the *ashi* point at the lateral side of patello-femoral joint. After disinfection, the doctor pushes the patella inward in order to expose the lateral femoral condyle. Then he inserts the needle and keeps the incision parallel to the vertical axis of the limb. At first, use the longitudinal dredging and stripping method, then shovel and debond once, and when the needle arrives at the surface of the bone withdraw the needle. Next push the patella outward so as to expose the medial femoral condyle and exert the same methods mentioned above.

Combined Treatment

1. Acupuncture with Collateral Bloodletting and Cupping Therapy:

Cheng Ting-xiu[23] combines the knee's five needle technique with collateral bloodletting and cupping therapy to treat this disease. The points include *hè dǐng*, ST 35, *nèi xī yǎn*, KI 10 (*yīn gǔ*) and BL 39. The three needle tips of *hè dǐng*, ST 35, *nèi xī yǎn* should form the shape of an equilateral triangle, while the other two needles, KI 10 and BL 39 should be inserted deeply into the periosteum and form a right angle. After the needling is over, apply the plum-blossom needle prick on BL 40 until the area bleeds slightly. Finish the treatment using cupping therapy over BL 40.

He Xiao-hong[24] inserts SP 10, *xī yǎn*, GB 34, ST 36, SP 9 and *ashi* points. After the qi arrives, he adds the electronic needle, bloodletting and cupping therapy on SP 10, BL 40 and the most tender spot.

2. Acupuncture with Medicinals:

Guo Fen-jin[25] uses acupuncture combined with medicinal therapy to treat

KOA. He inserts *nèi xī yǎn*, ST 35, GB 34 and ST 36 supplemented by SP 9, SP 10, ST 34 and *ashi* points with filiform needles rapidly, after the qi arrives, he adds an electronic acupuncture device on *nèi xī yǎn*, ST 35, GB 34, ST 36, SP 9 and SP 10 with a continuous wave form.

For the medicinal steam-wash therapy he often selects *xù duàn* (Radix Dipsaci), *wǔ jiā pí* (Cortex Acanthopanacis), *wēi líng xiān* (Radix et Rhizoma Clematidis), *rǔ xiāng* (Olibanum), *mò yào* (Myrrha), *niú xī* (Radix Achyranthis Bidentatae), *mù guā* (Fructus Chaenomelis), *dāng guī* (Radix Angelicae Sinensis), *bái sháo* (Radix Paeoniae Alba), *jī xuè téng* (Caulis Spatholobi), *chì sháo* (Radix Paeoniae Rubra), *hóng huā* (Flos Carthami), *bái zhǐ* (Radix Angelicae Dahuricae), *fáng fēng* (Radix Saposhnikoviae), *qiāng huó* (Rhizoma et Radix Notopterygii), *dú huó* (Radix Angelicae Pubescentis), *ài yè* (Folium Artemisiae Argyi).

Add *dà huáng* (Radix et Rhizoma Rhei) and *zé lán* (Herba Lycopi) for the patient with swelling, and add *zhì chuān wū* (Radix Aconiti Praeparata) and *zhì cǎo wū* (Radix Aconiti Kusnezoffii Praeparata) for the patient with painful joints aggravated by cloudy and rainy weather.

Wang Yong[26] inserts *ashi* points, *hè dǐng*, ST 36, BL 57. Others methods include penetrating ST 35 to *nèi xī yǎn* and GB 34 to SP 9.

The specific manipulation is as follows: perpendicularly insert the *ashi* point, and the other four points about 0.5 *cun* up, down, left and right towards the *ashi* point. Their needle tips should point to the middle; the point penetration method should be exerted with even manipulation; GB 34, BL 57 and ST 36 should be needled with the comprehensive manipulation using the mountain burning warming method.

The medicinals taken orally should comply with the principles of invigorating the blood and dissolving stasis, unblocking the collaterals and relieving pain, supplementing the liver and kidney, and strengthening sinew and bone. Some of the herbs used are *dāng guī* (Radix Angelicae Sinensis), *chuān xiōng* (Rhizoma Chuanxiong), *chì sháo* (Radix Paeoniae Rubra), *niú xī* (Radix Achyranthis Bidentatae), *wū shāo shé* (Zaocys), *dān shēn* (Radix et Rhizoma Salviae Miltiorrhizae), *sū mù* (Lignum Sappan), *bǔ gǔ zhī* (Fructus Psoraleae), *sāng jì shēng* (Herba Taxilli), *chuān shān jiă* (Squama Manitis), *huáng qí* (Radix Astragali) and *gān căo* (Radix et Rhizoma Glycyrrhizae). The steam wash therapy applied with medicinal compound formula can be used as well.

Zhou Yi-gui[27] uses warming needle moxibustion combined with intra-articular injection therapy to treat KOA. Warming needle moxibustion is often applied on ST 35, *nèi xī yăn*, ST 34, and GB 34. After the qi arrives, put the ignited moxa stick on the needle handle until the moxa burns out completely. Disinfect ST 35 and inject sodium hyaluronate directly into the articular cavity.

Problems and Prospects

Recently, there is a lot of clinical research focusing on the treatment of KOA. The primary treatment used in Western medicine are medical therapies or physical rehabilitation. Surgery is also employed on patients presenting with severe symptoms. Currently, TCM especially acupuncture, is widely pursued with interest for its unique advantages. Compared to other therapies, acupuncture has a lower cost, less pain, fewer side effects, and significant effectiveness which is easy to duplicate.

Acupuncture therapy includes filiform needling, collateral bloodletting, cupping, moxibustion, and acupotomy. All of them can be mutually supplemented or combined with oral medicinals, medicinal compress application, point injection therapy, catgut embedment and point application. All of these modalities main purpose is to dispel dampness, unblock the collaterals, invigorate the blood, move the qi and relieve the pain. Great achievements have been widely obtained in clinical practice.

A clinical research comparison of acupuncture treatment for KOA and domestic studies performed by Dr. Berman[28], chief counsel of the NIH, have revealed that there are some problems with these studies that need to be resolved:

(I) There is no unified standard of diagnosis or therapeutic effects in TCM. Additionally, there are some flaws in objective therapeutic evaluation due to different pattern differentiation and therapeutic evaluation.

(II) There are more clinical studies than animal experiments and the mechanical studies are not profound enough.

(III) Clinical research is not designed to comply with randomized control trials, and the sample capacities are small.

(IV) There is no united standard in acupuncture point selection. The needling intensity and frequency are lacking a quantifiable index. Furthermore, there is no persuasive objective index to evaluate the therapeutic effect, so further research and investigation is needed.

The therapeutic effect of acupuncture therapy is widely accepted, and it is often adopted as part of a comprehensive approach to therapy. But how do we know what an optimal treatment combination is during the different stages of KOA? What are the acupuncture laws of time and cycles? How do we apply EBM (Evidence Based Medicine) in acupuncture? And finally, how do we improve the standard of clinical research so that it is not just accepted, but respected, by Western medicine and its practitioners? All of these questions need to be addressed and answered in the near future.

References

1. Cong Xin. Summary of Current Situation and Problems of Knee Ostroarthritis Treated with Acupuncture (膝关节骨性关节炎针灸治疗现状和存在问题综述) [J]. Jilin Journal of Traditional Chinese Medicine (吉林中医药), 2005,25(7):58

2. Li Xiao-hao. Clinical Observation on 68 Cases of KOA treated with Single Insertion on LI 11 (单刺曲池穴治疗膝关节骨性关节炎68例) [J]. Chinese Journal of Clinical Rehabilitation (中国临床康复), 2004, 8(20): 4027-4031

3. Deng Bo-ying. Clinical Observation on 33 Cases of KOA's Pain Treated with Needling PC 6 (针刺内关穴治疗膝关节疼痛33例) [J]. Shandong Journal of Traditional Chinese Medicine (山东中医杂志), 2003, 22(8): 477

4. Wang Shu-qin. Clinical Observation on 49 Cases of KOA Treated with Acupuncture (针刺为主治疗骨性膝关节炎49例) [J]. Jilin Medical Journal (吉林医学), 2005, 26(6): 610-611

5. Zhu Qing-jun. KOA Treated with Abdominal Needle (腹针治疗膝骨关节炎) [J]. China's Naturopathy (中国民间疗法), 2009, 17(8): 12

6. Han Yue-Dong. 48 Cases of KOA Treated with Scalp Acupuncture and Body Acupuncture (头体针结合治疗膝关节骨性关节炎48例) [J]. Journal of Clinical Acupuncture and Moxibustion (针灸临床杂志), 2006, 22(1): 18

7. Chen Xiang-fei. 38 Cases of KOA Treated with Tug-of-war Needling (拔河针法治疗退行性膝关节炎38例) [J]. Journal of Clinical Acupuncture and Moxibustion (针灸临床杂志), 1998, 14(5): 30-31

8. Li Feng. Clinical Observation of KOA Treated with Valley Union Needling (合谷刺治疗膝骨性关节炎的临床观察) [J]. Gansu Journal of Traditional Chinese Medicine (甘肃中医), 2009, 22(5): 38

9. Qiu Ling. Clinical Observation of KOA Treated with Encircled Acupuncture (围针针刺法治疗膝骨关节炎的临床观察) [J]. Journal of Sichuan Traditional Medicine (四川中医), 2002, 20(11): 76

10. Hong Shu-chen. KOA Treated with Acupressure (中医指针综合治疗膝关节骨性关节炎) [J]. China Journal of Orthopedics and Traumatology (中国骨伤), 1998, 11(5): 23

11. Jin Jian-ming. KOA Treated with Massage and Muscle Training (推拿按摩结合肌力训练治疗膝骨关节炎) [J]. Chinese Journal of Rehabilitation (中国康复), 2006, 21(1): 42-43

12. Yang Zhi-qin. 34 Cases of KOA Treated with Moxibustion on ST 36 and Electronic Acupuncture (足三里灸配合电针治疗膝骨关节炎34例) [J]. Shanxi Journal of Traditional Chinese Medicine (陕西中医), 2009, 30(8): 1046-1047

13. Sun Kui. Clinical Observation of Primary Gnual Osteoarthritis of Liver-kidney Depletion Type by Aconite Cake-separated Moxibustion (隔附子饼灸治疗肝肾不足型膝骨关节炎的临床观察) [J].

Shanghai Journal of Acupuncture and Moxibustion (上海针灸杂志), 2008, 27(4): 9

14. Zhang Guang-li. 100 Cases of KOA Treated with Electronic Acupuncture and TDP (电针加TDP治疗膝关节炎100例) [J]. Journal of Anhui Traditional Chinese Medical College (安徽中医学院学报), 1996, 15(6): 39

15. Wu Ji-sheng. 36 Cases of KOA Treated with Electronic Acupuncture (电针治疗膝骨关节炎36例) [J]. Journal of Clinical Acupuncture and Moxibustion (针灸临床杂志), 2002, 18(4): 27

16. Wang Zong-jiang. KOA Treated with Point Injection (穴位注射治疗膝骨关节炎) [J]. Chinese Journal of Traditional Chinese Medicine (中国中医药杂志), 2005, 3(11): 988

17. He Cheng-qi. Clinical Research on the Point Injection and Motor Therapy to Treat Knee Osteoarthritis (穴位注射与运动疗法治疗膝骨关节炎的临床研究) [J]. Acupuncture Research (针刺研究), 2000, 25(3): 230

18. Wang Yun-xia. 82 Cases of KOA Treated with Point Injection and Heat Application (穴位注射加药热敷治疗骨性关节炎82例) [J]. Journal of External Therapy of Traditional Chinese Medicine (中医外治杂志), 1999, 8(1): 31

19. Li Ya-dong. 49 Cases of KOA Treated with Fire Needle (火针治疗膝关节骨性关节炎49例) [J]. Shanxi Journal of Traditional Chinese Medicine (山西中医), 2002, 18(3): 42

20. Dong Gui-fang. 145 Cases of KOA Treated with Fire Needle and Medicinal Application Compress (火针配合中药外敷治疗膝关节骨性关节炎145例) [J]. Journal of External Therapy of Traditional Chinese Medicine (中医外治杂志), 2001, 10(1): 7

21. Song Gao-feng. Senile KOA Treated with Needle-shaped Knife and Medicinal Application Compress (针灸刀配合中药外敷治疗老年退行性膝关节炎) [J]. Journal of External Therapy of Traditional Chinese Medicine (中医外治杂志), 1996, (1): 45

22. Li Jian-dong. 68 Cases of KOA Treated with Needle-shaped Knife (小针刀为主治疗髌股关节炎68例) [J]. Journal of Clinical Acupuncture and Moxibustion (针灸临床杂志), 1996, 12(2): 37

23. Cheng Ting-xiu. 48 Cases of KOA Treated with Five-knee Needles and Collateral Bloodletting and Cupping (膝五针加刺络拔罐治疗48例膝骨性关节炎患者) [J]. Chinese Journal of Physical Medicine And Rehabilitation (中华物理医学与康复杂志), 2002, 24(11): 670

24. He Xiao-hong. 46 Cases of Senile KOA Treated with Acupuncture, Collateral Bloodletting and Cupping (针刺加刺络拔罐治疗老年性膝骨性关节炎46例) [J]. Research of Traditional Chinese Medicine (中医药研究), 1994, (4): 50

25. Guo Fen-jin. Treating 52 Cases of Ostroarthritis in the Knee Joint with Needle Warming Through Moxibustion (温针灸治疗膝关节骨性关节炎52例) [J]. Chinese Journal of the Practical Traditional Chinese Medicine (中华实用中医药杂志), 2007, 20(23): 2048

26. Wang Yong. 129 Cases of KOA Treated with Acupuncture and Medicinal (针药并用治疗膝骨关节炎129例) [J]. Shandong Journal of Traditional Chinese Medicine (山东中医杂志), 2004, 23(9): 545-546

27. Zhou Yi-gui. 36 Cases of KOA Treated with Warming-needle Moxibustion and Articular Injection (温针配合关节腔内药物注射治疗膝关节骨性关节炎36例) [J]. China's Naturopathy (中国民间疗法), 2002, 10(7): 39

28. Guo Jia. The Gaps between Our Clinical Acupuncture Trials and the International Standards (对比分析针灸临床研究与国际接轨差距) [J]. Chinese Acupuncture & Moxibustion (中国针灸), 2004,24(1): 3-5

Index by General TCM Terms

Index by Acupuncture Point Names—Numerical Codes

Index by Acupuncture Point Names–*Pin Yin*

Index by Chinese Medicinals and Formulas

F

G

H

J

L

M

S

T

W

X

Z

图书在版编目（CIP）数据

针灸治疗膝骨关节炎＝Acupuncture and Moxibustion for Knee Osteoarthritis, A Clinical Series / 李万瑶，李万山主编．—北京：人民卫生出版社，2011. 4
（临床系列丛书）
ISBN 978-7-117-13973-1

Ⅰ. ①针… Ⅱ. ①李…②李… Ⅲ. ①膝关节－关节炎－针灸疗法－英文 Ⅳ. ①R246. 2

中国版本图书馆CIP数据核字（2011）第005619号

针灸治疗膝骨关节炎——临床系列丛书（英文）

主　　编：李万瑶　李万山
出版发行：人民卫生出版社（中继线+8610－5978－7399）
地　　址：中国北京市朝阳区潘家园南里19号
世界医药图书大厦B座
邮　　编：100021
网　　址：http: //www.pmph.com
E - mail：pmph @ pmph.com
发　　行：pmphsales@gmail.com
购书热线：+8610－5978 7399/5978 7338（电话及传真）
开　　本：850×1168　1/32
版　　次：2011 年 4 月第 1 版　2011 年 4 月第 1 版第 1 次印刷
标准书号：ISBN 978－7－117－13973－1/R・13974